Essentials of Engineering Design Graphics

A Modular Approach to Graphics for Engineers

Gerald E. Vinson

Texas A&M University

KENDALL/HUNT PUBLISHING COMPANY
4050 Westmark Drive   Dubuque, Iowa 52002

Copyright © 2002 by Kendall/Hunt Publishing Company

ISBN 0-7872-9386-5

All rights reserved. No part of this publication may be reproduced, stored in a retrieval system, or transmitted, in any form or by any means, electronic, mechanical, photocopying, recording, or otherwise, without the prior written permission of the copyright owner.

Printed in the United States of America

10 9 8 7 6 5 4 3 2 1

Dedication

This book is dedicated to my high school sweetheart, my wife, and my best friend . . . who all happen to be the same person. With my deepest gratitude for all of your assistance, inspiration, patience, and encouragement throughout this project.

Contents

Introduction vii
Why Engineering Design Graphics vii
Modular Instructional Materials viii
Preparing for Work in Industry viii

1 Communicating Across Time—Lines & Lettering 3
Freehand Lettering 4
10 Basic Rules for Lettering Notes and Dimensions 5
Some Tips From the Pros 9
Tools of the Trade 10
Practice Quiz: Lettering and Basic Lines 15

2 Sketching for Engineers 17
Effectiveness of Sketches 19
Basic Elements of Sketches 20
Building Blocks for Sketches 21
Line Sketching Techniques 23
Circle Sketching Techniques 25
Some Tips from the Pros 28
Sketching Summary 28
Pictorial Sketches 29
Axonometric Pictorials 30
Oblique Pictorials 36
Perspective Pictorials 39
Practice Sketching 43
Practice Quiz: Sketching and Pictorials 45

3 Detailed Sketches 47
Using Different Linetypes 51
Getting Started 55
Proper Definitions 57
Reviewing the Basics 57
Orthographic Drawing Summary 59
Practice Quiz: Multi-View Drawings 61

4 Specifications and Dimensioning 63
Rules for Dimensioning 65
Placement of Dimensions 74

Table Driven Dimensions 75
Dimensioning Standard Parts 76
Practice Quiz: Dimensioning 87
Review of Dimensioning Rules 88

5 Screw Threads and Fasteners 91
Basic Definition 92
Thread Terminology 93
Thread Notations in U.S.A. 93
Metric Thread Notation 94
Thread Symbols 94
Internal Threads 96
Square and Acme Threads 96
Types of Bolt Heads 96
Practice Quiz: Threads and Fasteners 101

6 Special Types of Views 103
Sectioned Views 104
Broken Out Section 104
Full Section 107
Half Section 109
Revolved Section 110
Removed Section 111
Offset Section 112
General Rules for Sectioned Lines 113
Auxiliary Views 116
Descriptive Geometry Concepts 118
Practice Quiz: Section and Auxiliary Views 121

7 Tolerances in Design 123
Expression of Tolerances 124
Design for Mating Parts 126
Types of Fits 127
Summary of Tolerancing 131
Geometric Tolerance 133
Practice Quiz: Tolerances in Design 141
Practical Exercises 142

8 Working Drawings 145
Layout and Analysis Drawings 145
Details and Specifications 145
Assembly Drawings 145
Sheet Layouts 146
Title Blocks 150
Parts Lists and Revision History 151
Assembly Views 151
Practice Quiz: Working Drawings 157

9 Engineering Design 159
The Design Process 159
Steps of Engineering Design 160
Design Teams 161
Logical Steps from Design to Solution 162
Patent Resources 163
Sample Design Report 167
Practice Quiz: Engineering Design 177
Design Project Ideas 178

10 Computer Aided Drafting 181
Historical Concepts 181
Common Software Concepts 181
Specifications to Solids 183
Rapid Prototypes 183
Practice Quiz: Computer Aided Drafting 185

Appendix A 187
Appendix B 205
Index 211

Introduction

Why Engineering Design Graphics

This book is intended to provide you, the beginning engineer, with the basic information needed to complete a first course in Engineering Design Graphics (EDG). There are numerous other textbooks on the subject, which go into volumes of minute detail, contain 800–1000 pages, and are very expensive. In order to reduce costs, conserve paper, and provide just the "Essentials" necessary for a first course, many of the older, less used topics, have been minimized or omitted. At the same time, many figures have been provided at full scale to clarify key topics. As you enter the engineering profession, you will most likely find that the company you work for has created its own standards for graphic style and format. Until that time, this book will serve as a reference for the most commonly used EDG rules based on the graphic standards of the American National Standards Institute (ANSI).

Courtesy of MSCUA, University of Washington Libraries. Negative nos. UW 21679, UW 21422.

"Often, the differences between design concept and design reality have dramatic results."

Modular Instructional Materials

A more flexible, modular approach is being used with this text that includes two additional modules. One module is an engineering design graphics workbook to provide "hands on" practice for creating engineering sketches and finished drawings on worksheets with pencil and/or computer. The other module will provide you with the information required to create drawings using a computer aided drafting (CAD) program such as AutoCAD. These three modules should provide all the information you will need for a one or two semester course in engineering graphics. The only things you will need to supplement this modular approach are a few good drafting pencils, an eraser, some creativity, and the willingness to complete the assignments on time and within specifications.

The instructional topics are presented in a manner which ramps the student up to the desired level of competence, one block at a time. Many full sized examples are provided for your reference throughout this text. A typical block of instruction would include a brief lecture followed by some sketching exercises from the workbook. Once these sketches are graded or critiqued by team members, they are converted into computer drawings, printed out, and submitted for a grade. Normally, the engineering sketches are attached to the CAD drawings and the entire set is graded and returned for future reference.

Preparing for Work in Industry

In industry, the same type of CAD drawings would be reviewed for accuracy, approved by an engineer, and forwarded for the patent process or production. The cycle would then continue as the product is built, tested, sold, supported, and continually improved over its product life cycle. Engineering notes, accurate sketches, and computer files are required to support and defend this production cycle. This book will teach the tools of the trade and skills necessary to facilitate this concurrent design/manufacturing cycle.

Design is the key to the entire product development, and it is a key element of this text. Good design can make the difference between a robust product that is functional and profitable and one that is a failure. The principles of working in design teams and the design process will be heavily emphasized in this text. No other element of engineering has the potential for greater rewards.

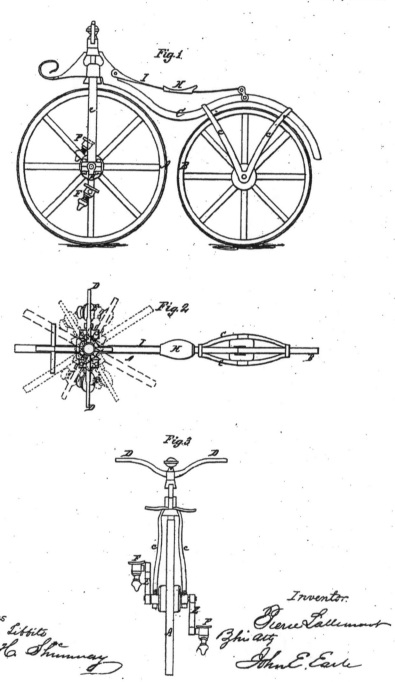

Communicating Across Time—Lines & Lettering

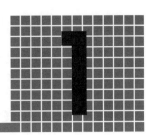

Figure 1.1

Early forms of graphic communications have been discovered in prehistoric caves and archeological digs that date back farther than 5000 B.C. These consisted mainly of pictures on cave walls and writing on vases and clay tablets. Early hieroglyphics did not convey sounds, but rather told stories with pictures . . . much the same as we do with engineering design graphics today. **(Figure 1.1)** It was left up to the Phoenicians to create the first alphabet that represented sounds in about 1500 B.C. Ever wonder where the word "phonics" came from? Their alphabet contained only 22 letters that were mostly consonants. Later, the Greeks took fifteen of these letters and added some of their own for a total of twenty-four characters. This became the Latin alphabet around 700 B.C. It was called an alphabet after the names of the first two letters . . . Alpha and Beta. When the Romans conquered the ancient Greeks, they kept eighteen of the Greek letters and added seven more, including some characters, to form their own alphabet. After the fall of the Roman Empire, the Anglo-Saxons adopted all twenty-five of the Roman characters and added two of their own. Later, they dropped the "thorn" (which was "th") yielding a total of twenty-six letters. The final result is the alphabet we use today. Many of the original characters have descended relatively unchanged after thousands of years.

The style and form of these early letters varied with the penmanship of the individual scribes. **(Figure 1.2)** Real uniformity of alphabets did not evolve until a German, Johann Guttenberg, invented the printing press with moveable metal types in

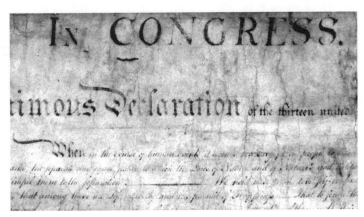

Figure 1.2

1452. (**Figure 1.3**) At last there was a degree of consistency for printing that the public could use as a model for its own printing and handwriting.

Freehand Lettering

Sizes of parts and notations about materials and construction have been placed on engineering drawings for just about anything humans have made for thousands of years. For the past 100 years or so, there have been concerted efforts to make the drawing and hand lettering styles consistent in size and form. In 1893, an American named C.W. Reinhardt modified the Gothic alphabet to omit serifs and thick lines for hand lettering. He called this new style "single-stroke gothic." Mr. Reinhardt published an article about his new lettering style in the 1893 edition of "Engineering News." In the article, he claimed that this new style of hand lettering was clearer and faster to apply than traditional styles

Courtesy of the Deutsches Museum.

Figure 1.3

of that day. He recommended using all capital, vertical letters as shown in **Figure 1.4**. Many engineering companies must have agreed with this new style, as it was widely used on engineering drawings at the turn of the century. In 1935, the American National Standards Institute (ANSI . . . commonly referred to as "Ann-see") adopted this style for their recommended form of hand lettering. The ANSI reviews and updates their standards on a regular basis and when the lettering standard "Y14" was revisited in 1992 it was left relatively intact. A comparison between an old and the newer style is shown in **Figure 1.5**. The formation of the individual letters that has been traditionally used is shown in **Figure 1.6**. Some important rules for lettering are listed below. It is critical to learn and practice the use of the rules on all engineering drawings.

A B C D E F G H I J K L M
N O P Q R S T U V W X Y Z
1 2 3 4 5 6 7 8 9 10

Figure 1.4 Single Stroke Vertical Gothic alphabet and numerals.

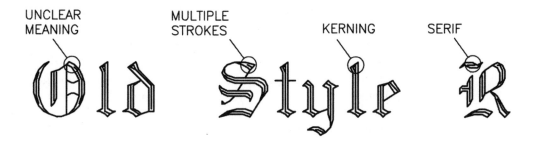

The style below is much easier to read and understand. It also maintains clarity with copies, faxes, and drastic reductions in size.

SINGLE STROKE IS CLEAR & CONCISE

SINGLE STROKE IS CLEAR & CONCISE
SINGLE STROKE IS CLEAR & CONCISE
SINGLE STROKE IS CLEAR & CONCISE AND HOLDS UP TO REDUCTIONS VERY WELL.
SINGLE STROKE IS CLEAR & CONCISE AND HOLDS UP TO REDUCTIONS IN SIZE VERY WELL.
SINGLE STROKE IS CLEAR & CONCISE AND HOLDS UP TO REDUCTIONS IN SIZE VERY WELL.

Figure 1.5 Single Stroke Vertical Gothic compared with Old English. The single line elements can be drastically reduced in size and still be legible.

10 Basic Rules for Lettering Notes and Dimensions are as Follows:

1. Use only Single Stroke Gothic lettering.
2. Lettering shall be all capitals.
3. Finished lettering height should be .125 inches or 3mm.
4. Titles should be ¹/₄" or 6mm in height.
5. Finished height for common fractions shall be .25 inches or 6mm.
6. Vertical spacing between lines of lettering shall be no smaller than one half of the text height.
7. Common fractions are twice the height of normal letters.
8. Guidelines should be very light (almost invisible). Use a 4H or harder to draw guidelines.
9. If inclined letters are required, they should make an angle of 68° with the guidelines.
10. Spacing between letters in the same word should be based on volume instead of a fixed linear distance.

These rules should be practiced on all hand drawings and the same size and style should be duplicated as nearly as possible on computer-generated drawings. It is important to note that this is not just an American standard, but in this age of global engineering and manufacturing, it has been adopted by most industrialized nations. The necessity for clear, concise, and easily read documents is heightened as we depend more and more on scanned documents, faxes, and attachments to emails to transmit our designs and specifications worldwide. When numerals are used on drawings, the international community generally prefers that they look like those shown in **Figure 1.7**.

The basic concept for making quick letters is that the last stroke you make on a letter should move your fingers toward the next letter you are going to print.

Figure 1.6 Different strokes for different folks? The stroke sequences for constructing letters shown above was proven to be very efficient by the old masters. You may adapt it to a sequence that is faster or easier for your individual needs.

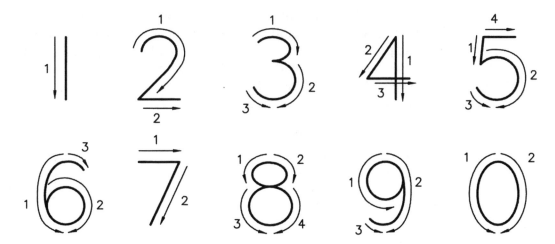

Figure 1.7 Single Stroke Gothic numerals are shown above. Since engineering design is driven by numerically controlled dimensions, it is important to master the quick and clear printing of each number. In order to insure accuracy and the clear communication of your design intent, do whatever it takes to make your printing identical to these.

Letter Spacing Rules

1. The space between hand-lettered words should be equal to the width of the letter "O." (**Figure 1.8**) The space between individual letters in a word should be based on the apparent volume instead of a uniform linear spacing. Linear spacing makes some letters look too close to each other and others look too far apart. (**Figure 1.9**)
2. The vertical space between hand-lettered sentences should be no less than one half the text height and no greater than one letter height.
3. Common fractions have horizontal fraction bars that are slightly wider than the numerals they separate.
4. The total height of common fractions is two letter heights. (**Figure 1.10**)
5. The fraction numerals should be about 20% shorter than the text for whole numbers.
6. Common fraction numerals should not touch the fraction bar.

Figure 1.8 Spacing between words should be equal to the width of the letter "O." Vertical spacing between sentences should not be any closer than one half of the text height.

Practice, practice, and more practice is the only way to develop neat lettering skills. Engineering students should make use of upper case, single stroke, gothic lettering at every opportunity. Use it for taking notes in all of your classes and even for writing letters to family and friends. You will find that speed, clarity, and consistent form become automatic after repeated use.

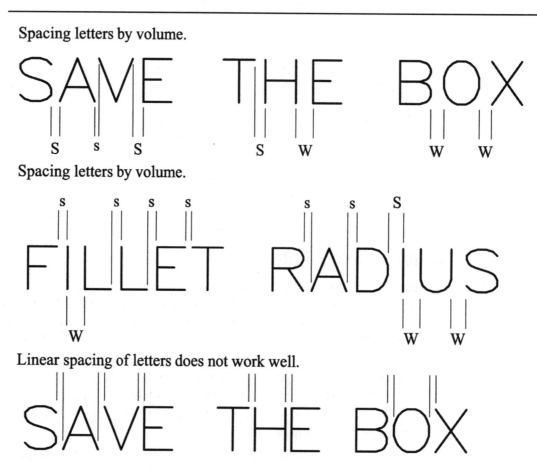

Figure 1.9 Spacing between letters can be tricky when it is done freehand. The spacing should be treated as a volume instead of a linear distance. Some letter combinations look better with wide "W" spaces while others need a smaller "S" space between letters.

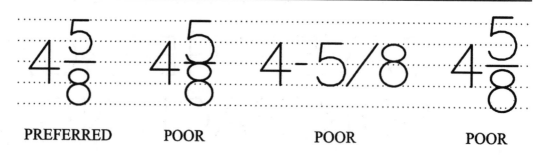

Figure 1.10 Common fractions should not be taller than twice the letter height. Numerals should not touch the horizontal line so their height must be slightly shorter than normal text.

Some Tips from the Pros

TIP 1: Learn to make the strokes for "**L**," "**C**," and a backward "**C**" quickly and consistently. The "**L**" stroke is used to form the vertical and bottoms of the letters **B, D, L,** and **E** with a quick down and over stroke. Starting the **B** and **D** with this "90° heel stroke" will keep them from being mistaken for an **8** or a **0** if the bottom left corner assumes a lazy rounded form. The forward "**C**" and backward "**C**" are actually half circles that are used to make all of the letters that have curved elements. These are the letters **B, C, D, G, O, P, Q, R, S,** and **U**. The **C** and **G** make use of the forward stroke, the **S, O,** and **Q** use both strokes, and the letters using the backward "**C**" stroke are the **B, D, P,** and **R**.

TIP 2: Always use a soft pencil lead with a slightly rounded point (intentionally dulled) for lettering. The drawing pencil grades of "H and 2H" are recommended for consistent darkness and resistance to smudging. Even though the "H" lead is about the equivalent softness of a typical "no. 2" pencil, it is not recommended to use the cheaper numbered varieties of leads. Drafting pencil manufacturers guarantee the consistency of lead hardness to match those already on a finished drawing, year after year. A "no. 2" pencil will often vary in hardness even when made by the same company, and often produce fuzzy lines, which do not erase cleanly.

TIP 3: When writing with freehand lettering, it is also a good idea to turn the writing paper to a comfortable angle as you would when doing cursive writing. Always use a grid paper or draw guidelines to control the height of your freehand text. Letter height templates and devices to quickly draw very light guidelines like the Ames lettering guide are available from drafting and office supply businesses. Even beginner's lettering can look somewhat professional when the letters are all the same height and in straight lines.

TIP 4: Place your entire forearm on the table for best support and hold the pencil so it makes an angle of approximately 60° with the paper. Rotate the pencil in your grasp every four or five letters to keep from wearing a flat spot on your lead. This rotation should keep a proper, slightly rounded point on your lead. Even mechanical lead holders with the .5mm or .7mm diameter leads require this kind of rotation in order to prevent making letters that have thick and thin elements.

TIP 5: When you are lettering on a large drawing or creating multiple lines of text, it is a good idea to place a sheet of clean paper under the heel of your hand where it rests on the drawing. This will keep the oil or moisture in your skin from smudging the text or sketches already on the page. Work from top to bottom and try to keep the drawing surface extremely clean. Even fingerprints will cause the lead to smudge easily and repel India ink.

TIP 6: Whenever possible, wait and apply all of your lettering at the same time, preferably after all of the line work is finished. The resulting letters will be more consistent in form and darkness when applied in one setting. In all freehand printing, the finished lettering must appear black, and there is only one shade of black. Gray, fuzzy shades of lettering will not make clear copies or legible faxes and can cause errors and costly mistakes in manufacturing.

Clear Communication Requires Quality Drawings

Engineering graphics are created with lines, arcs, and circles. Special types of lines are used for making engineering drawings. Frequently called "the alphabet of lines" or "linetypes," they are divided into two broad categories: **Continuous** and **Noncontinuous**. In the past, there were thin, medium, and thick widths in both categories, but today most engineers use only thick and thin lines. This practice has evolved with the development of computer graphics and the limited pen settings of early plotting devices. The commonly used plotter pen widths were 0.7mm for the thick lines and 0.3mm for thin lines. Current technology for laser printers allows for line weights to be set to practically any desired width and the ANSI recommends 0.6mm for thick and 0.3mm for thin lines. Extremely large or very small drawings may require thicker or thinner weights depending on the desired final plot scale, i.e. a large poster or postage stamp. Additionally, many companies have their own drafting room standards and design templates which already have their line thickness specified.

Tools of the Trade

In addition to hand printed text, drawings include the use of lines, arcs, and circles. Hand sketched or precision drawings can be drawn with properly sharpened wooden pencils or mechanical lead holders with lead diameters of 0.7mm for the thick lines and 0.3 mm for thin. These provide a degree of uniformity as compared to drafting pencils that get dull with use and draw wider, fuzzier lines. **Figure 1.11** shows the typical types of pencil lead points used for sketching. The chisel or wedge is easily sharpened in the field and will keep a sharp edge as the corners are alternated as they become worn down until a sharp point is left in the center. Some of the mechanical pencils with larger diameter leads have built in lead pointers. The micro point mechanical pencils do not require sharpening but sometimes require two or three passes to achieve a bold linetype.

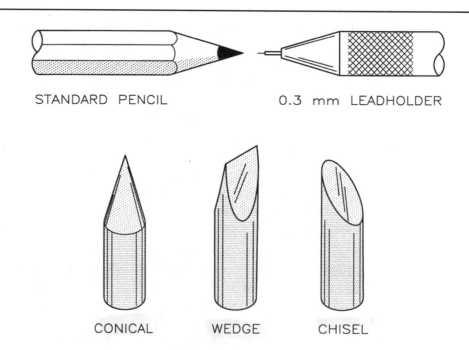

Figure 1.11 Engineering drawings can be made with any of these types of lead points. The micromatic type pens do not require sharpening and produce better consistency of lines and lettering.

In addition to the three types of points to be chosen, there are also a variety of pencil lead hardnesses to be considered that range from the **softest 7B** to the **hardest 9H**. **(Figure 1.12)** The hardness of the lead determines how long a pencil lead will hold its shape. Hard leads can be used for very thin lines where softer leads should be used for thicker lines and lettering. Leads are manufactured by blending a mixture of graphite and clay. They are extruded to size and then baked to make them solidify. Harder leads have more clay while softer leads have a greater ratio of graphite. Since graphite is basically a lubricant, leads containing higher amounts have a tendency to smudge and are difficult to erase cleanly. Artists and illustrators make use of soft lead grades marked "**B**" more often for rendering where it is desirable technique to smudge and blend. Engineering designers, however, generally keep about six of the harder leads available for use. The three that seem to be used the most often are the **H, 2H, and 4H** and these are all you will really need to get started. Most designers use "H" for lettering and shading, "2H" for thick lines, and the "4H" for thin lines, guidelines, and light sketches. What works for them may not work for you. Each individual needs to experiment and find which grade of lead works best for him or her. The individual pressure applied to the paper and the angle at which the pencil is held

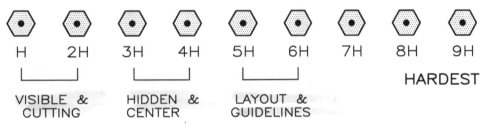

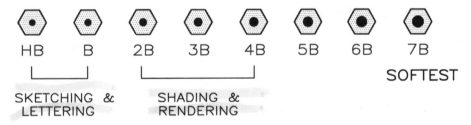

Figure 1.12 Various grades of drawing leads are available for engineering drawings and sketches. Many variables such as humidity, type of paper, and force applied to the paper will determine which one works best for you.

affects text and line quality on the drawing. Other variables affecting good pencil work include, but are not limited to, the quality of drafting paper and the humidity present in the room. These two alone can make the same pencil print darker on some days than on others.

The commonly used linetypes are illustrated in **Figure 1.13**. Note that all lines for engineering drawings must be black. Whether thick or thin, they must produce sharp, readable copies and faxes so lighter colors are not recommended. The linetypes fall into two categories known as **continuous** and **noncontinuous**.

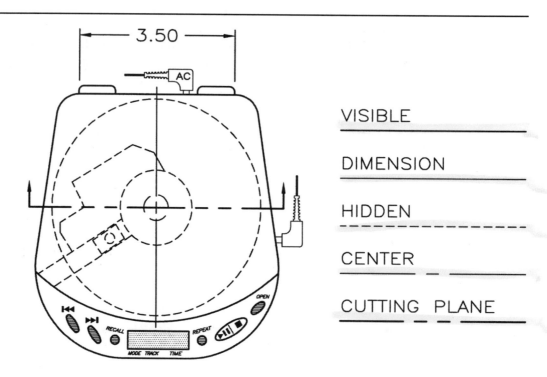

Figure 1.13 Examples of various types of continuous and noncontinuous linetypes required on a engineering drawing.

Continuous

Visible lines, sometimes called **"object lines,"** are drawn around the outside edges of visible objects and indicate sharp bends in planar surfaces. Other continuous lines are:

- **Dimension lines**—used to indicate sizes of features
- **Extension lines**—used to show where the dimension originates
- **Some crosshatch lines**—used to indicate a parts material

Noncontinuous

These lines are made up of combinations of lines and gaps at regular intervals.

- ❒ **Center**—indicates the center of arcs and circles
- ❒ **Hidden**—used to indicate features whose outline is blocked from viewing
- ❒ **Cutting plane**—indicates where a theoretical knife slices through the part
- ❒ **Some crosshatch lines**—patterns such as brass use hidden lines
- ❒ **Phantom**—used to "ghost in" alternate positions and mating parts

The size of the gaps will vary with the scale of the drawing, but for most full size layouts dashes for hidden lines are about .125" long and the gaps are about .10" long. **(Figure 1.14)** These sizes are approximate because designers using pencils rely on "eyeball spacing" to estimate the lengths. Long or short, consistency is the main objective, so you should be consistent over the entire set of drawings that describe your design. When your sketches are finalized with a CAD system, the software will most likely have a variable for scaling linetypes to have long or short dashes. This option affords consistency and is typically set up as a template file for repeated use.

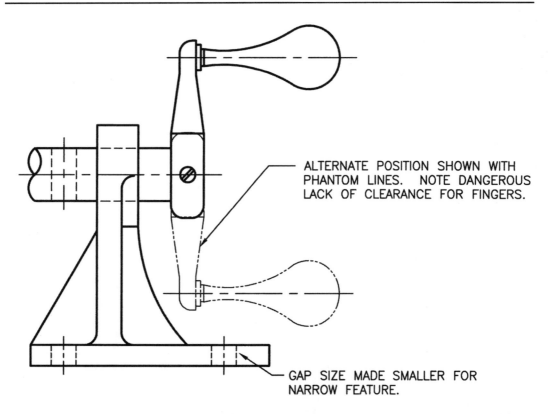

Figure 1.14 Phantom lines are used to show the alternate positions of moving parts for functional analysis. Notice that the scale of the hidden and center lines varies to match the thickness of the feature.

CHAPTER 1—Practice Quiz: Lettering and Basic Lines

_____ 1. Hand lettering by engineers is most likely to contain _____ .
A) Serifs   B) Single strokes   C) Kerning   D) Double-wide strokes   E) None of these

_____ 2. The style of hand lettering used by engineers today was first proposed in the late 1800's by _____ .
A) Arial   B) Courier   C) Serif   D) Reinhardt   E) None of these

_____ 3. In most applications, all text that appears on engineering drawings and sketches should be _____ .
A) Black   B) .125" tall   C) Vertical   D) Uniform in style   E) All of these

_____ 4. The primary goal of text and graphics on engineering drawing is _____ .
A) Being brief   B) Clear communication   C) Ease of copying   D) Attractive layout   E) None of these

_____ 5. Among the following pencil leads, the one that would be recommended for hand lettering is _____ .
A) 4H   B) 4B   C) H   D) 6H   E) All of these

_____ 6. The pencil lead listed below best suited for drawing thin guidelines is _____ .
A) 4H   B) 4B   C) H   D) 2H   E) None of these

_____ 7. The leads used for engineering drawings are manufactured to be harder with the addition of _____ .
A) Iron ferrite   B) Lead   C) Charcoal   D) Clay   E) None of these

_____ 8. The leads used for engineering drawings are manufactured to be softer with the addition of _____ .
A) Wax   B) Graphite   C) Charcoal   D) Black ink   E) None of these

_____ 9. Both CAD and hand lettering of specifications on engineering drawings should always have _____ .
A) The first letter of each word capitalized   B) All letters in uppercase
C) Only titles in uppercase   D) Fractions italicized   E) None of these

_____ 10. According to the alphabet of lines, the standard linetype used to show the silhouette of an object on an engineering drawing would be called a _____ .
A) Outline   B) Dotted line   C) Perimeter line   D) Visible line   E) None of these

_____ 11. A line that consists of a series of alternating long and short dashes is called a _____ .
A) Broken line   B) Hidden line   C) Center line   D) Dashed line   E) None of these

_____ 12. The type of line that is used to show features that are invisible to the viewer is the _____ .
A) Invisible line   B) Hidden line   C) Ghost line   D) Dotted line   E) All of these

_____ 13. Linetypes used by both CAD and hand drawings fall into two main categories which are _____ .
A) Black and gray   B) Continuous and noncontinuous   C) Solid or dotted
D) Sketch and detailed   E) None of these

_____ 14. The recommended vertical space between hand lettered sentences in the same paragraph should never be less than _____ .
A) .25"   B) One half the text height   C) One text height   D) .125"   E) None of these

Sketching for Engineers

The day is coming in the not too distant future when we will be able to talk to a handheld computer and describe our design ideas. This little talk will document the specifications, generate drawings, and verify the design intent. Until that day however, we're faced with having to record our ideas quickly and efficiently so they will not be forgotten. The traditional method is to make quick, engineering sketches with pencil and paper. Engineering sketches differ from most artistic sketches by the fact that they are line drawings in black and white with numerous notations. Conceptual sketches like the one shown in **Figure 2.1** are intended to capture the early design ideas. In order to verify ownership of the design, the sketch should always be signed and dated. Other precautions that should be observed for conceptual patent sketches are discussed in Chapter 9.

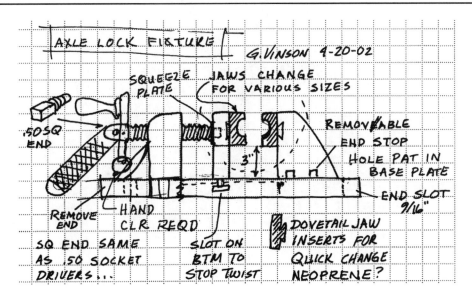

Figure 2.1 Conceptual sketches capture the engineer's design intent. Quickly capturing the design concepts is more important than accuracy at this early stage of the design process.

Generally, these engineering sketches will be composed of outlines of each part or mechanism in an assembly. Artists are more concerned with making the artwork look real, whereas engineers are concerned with clearly communicating the design intent, and capturing the details necessary for production or modification.

Most often, engineering sketches are made in a 2-D format, but pictorial sketches may be included for clarity and understanding. Engineers seem to prefer pictorials that utilize parallel drawing planes such as axonometrics and obliques, but sometimes use perspectives for realism as shown in **Figure 2.2**.

Sketches are particularly useful in the conceptual or brainstorming stage of product development. The main emphasis for engineering drawings and sketches must always be on **communicating the engineer's design intent** to those who make the final drawings and prototypes. After testing the prototype, parts are often changed and require revised and updated drawings . . . so the design improvement loop begins all over again.

Early sketches and drawings should be archived for easy reference in case the design needs to be reworked. This is especially important if a product failure results in a lawsuit. The engineer who can document that the design was altered from the original specifications should not be held culpable. Experienced engineers recommend that you cover yourself with backups and copies. The thicker the papers, the fewer the teeth can bite through it . . . so **cover yourself always** (or "**CYA**" as practicing engineers say) with sketches, detailed drawings, notes, and specifications. The trend toward concurrent engineering and working in teams makes the need for document control and access even more critical.

Effectiveness of Sketches

Good, concise, and accurate engineering sketches are the key to producing accurate computer drawings. If the CAD technician has to contact the engineer for clarification of the engineer's sketches or notes, then the engineer's sketches were not effective. At the very least, this can result in lost work time in the design process, or at worst, could result in the production of inaccurate, faulty parts.

Engineers need to be very good sketch artists and must print all communications very neatly. There are three main phases of becoming a good engineering sketch artist. They are:

1. **Practice**
2. **Practice some more**
3. **Practice a whole lot more**

No reasonable person would consider going into competition in athletic or intellectual games without practice. Sketching is the same way. The more you practice the faster and better you become. Sorry, but there is no other way. In time you'll get proficient with CAD and can work more independently but you will find it is still

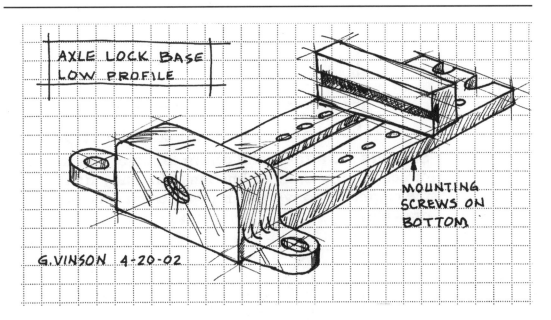

Figure 2.2 Pictorial sketches greatly help to communicate the early design. These are sometimes done in real time during presentations and brainstorming sessions.

easier to input data into CAD drawings when you have a sketch to reference. **(Figure 2.3)**

Basic Elements of Sketches

The most primary elements of sketching are lines and arcs. There are two schools of practice for drawing straight lines. Each of these methods requires that you grip the sketch pencil (H or 2H) an inch or two farther back than when you are using it for instrument drawings or normal text applications. A "4H" pencil is perfect for construction lines and light layout work. Because of the hardness of this lead it will

Freehand sketches are the fastest way to capture an idea or relay instructions to the CAD technicians. Light construction lines and corner over-runs are acceptable, as speed is more important than accuracy in this phase.

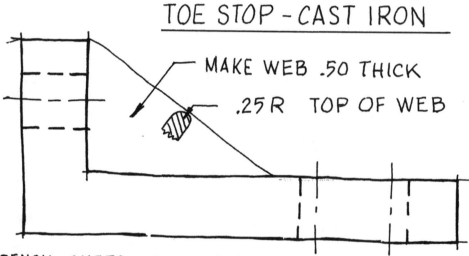

Sketches are refined and drawn to scale with instruments or sent directly to the CAD technicians. In many fields of engineering the engineers convert their own design sketches to computer drawings.

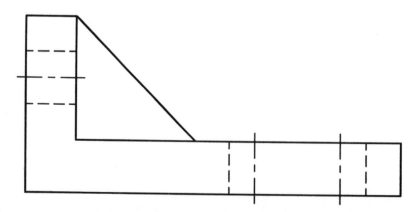

Figure 2.3 Sketches play an important role in the design/manufacturing process.

hold its sharpness four or five times longer than an H or 2H lead and resists smudging much better. Pencils should have sharp points for light construction layouts, and slightly dulled points for drawing the wider, visible outlines. These visible outlines should be at least twice as wide and bold as construction lines. Visible lines should always be black . . . and remember, there is only one shade of black.

It is easier to sketch if your sketchpad is positioned on a flat surface and rotated to be at the same angle you would use if you were writing a letter. **(Figure 2.4)** Good quality drawing paper holds up better to erasures and is more durable with repeated folding or rolling. Professionals frequently use acid free, 100% rag paper for their drawing archives. This paper costs considerably more initially, but pays for itself over time. Leonardo da Vinci, who is probably the most prolific sketch artist of all time, made over 5000 sketches in the 14th century that are still displayed in museums today. Not bad, especially when you consider that he was using homemade ink in goose quills to sketch on leather sheets.

Figure 2.4

For best results, you should hold the pencil at an angle of about 60° to the sketchpad. The soft grades of leads used for sketching will become dull very quickly. For this reason, you should rotate the pencil about one fourth of a turn after each long stroke and the lead will wear evenly on each quadrant, thus staying relatively sharp. Failure to rotate the pencil will result in a flat side or chisel shaped lead, which could result in making fuzzy, thick and thin lines as the path of the pencil changes directions. **(Figure 2.5)**

Building Blocks for Sketches

After you get pretty good at sketching lines and circles you will be ready to master combining them into the three basic building blocks of sketching. These basic building blocks are the **square, circle,** and **triangle**. **(Figure 2.6)** They may be combined to create sketches of both natural and manmade objects. Notice in the sketch of the "Jeepster" that these shapes are all around us in our everyday lives. In time, you will learn to break objects down into these basics and you will be able to create sketches of anything you desire. The old Texas church shown in **Figure 2.7** would be fun and challenging to sketch. Consider that before it was built, the designer had to create

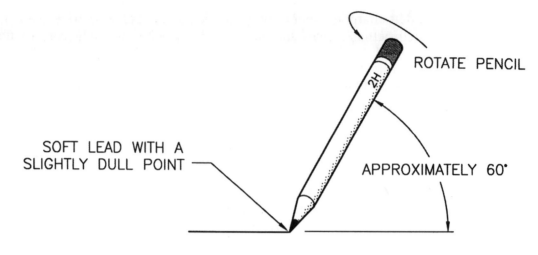

Figure 2.5

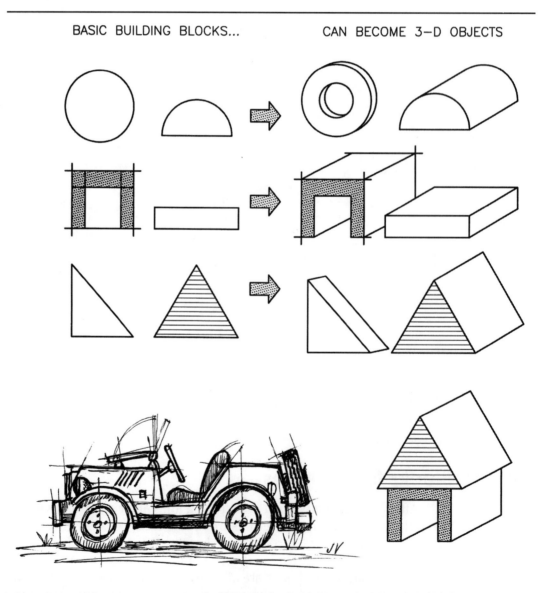

Figure 2.6 The basic building blocks for engineering sketches are the **circle, square and triangle**. These easily convert to extrusions as **boxes, cylinders, or wedges** for three dimensional representations.

Figure 2.7

many sketches to sell the design to the church's congregation. These combinations of basic shapes existed only in the designer's mind until they were shared with others as sketches. Just imagine the impossible task of trying to describe this church design without sketches. When you learn to master drawing these basic shapes, you can sketch anything. These shapes can be extruded to become **boxes**, **wedges**, **cylinders**, and **cones** which are the foundations for pictorial sketches. When pictorials containing circles are sketched, the circular features often become ellipses, so it is a good idea to learn how to draw smooth ellipses as well as circles. These "squashed circles" are discussed in greater detail later in this chapter.

Line Sketching Techniques

The first technique is to place two marks on the paper to represent each end of the line **(Figure 2.8)**. Put the pencil on one point and try to pull the pencil to the other endpoint. The trick is to keep your eye on the second point and make a quick, light stroke. Once you are satisfied with the straightness of your construction line it is easy to trace it with more pressure to make it darker. Pure sketching does not use any type of straight edge or template, and does not require that you erase the light construction lines. Practice by sketching five or ten parallel lines at a time. Vary the angles and repeat the process until you can hit the target point consistently with your light construction lines. Old draftsmen will tell you that if you **"hurry up"** they will **"look alike,"** so always lay down your light construction lines as quickly as possible. **Figure 2.9** shows the setup for multiple lines.

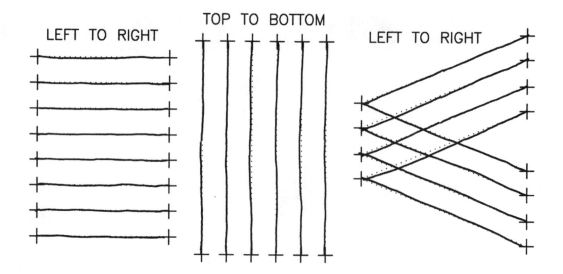

Figure 2.8 Practice sweeping your sketch pencil to the target. Practice at various angles and strive to make quick construction lines that are very light. Longer lines may require a series of shorter segments.

The second method of sketching lines involves making a series of very light, short dashes instead of one long stroke. **Figure 2.10** shows this dashed line technique. The light construction dashes are refined or replaced where they do not fit the intended path, and the good ones are traced with a dark line that follows the desired path. This technique seems to work best for the sketching of longer lines, circles, and arcs. **(Figure 2.11)**

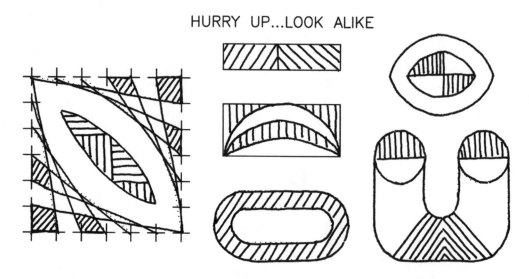

Figure 2.9 Another method of sketching lines is to connect a series of short dashes as the construction guide. This also works when sketching arcs and circles.

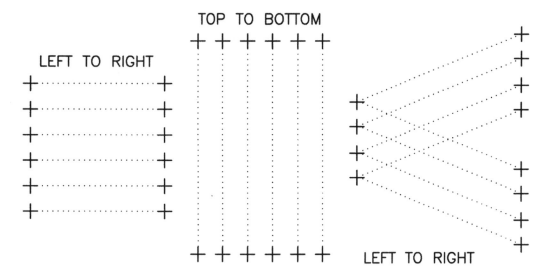

Figure 2.10 Practice by following the dotted lines to the targets. Move the sheet around to find your most efficient sketch angle and pencil stroke.

Circle Sketching Techniques

Beginners seem to have the most trouble sketching circles. These four easy steps will help you to sketch rounder circles:

- If the circle is smaller than an eighth of an inch, just make the roundest letter "O" that you can with one continuous loop that starts and ends at the top. With practice they will appear very round in both your sketches and your text.
- When the circle size is between an eighth and a half of an inch, create it with two semicircles. Make the first pencil stroke as you would the letter "C" and finish it off with a second stroke like you would make the curved part of the letter "D." **(Figure 2.12)**

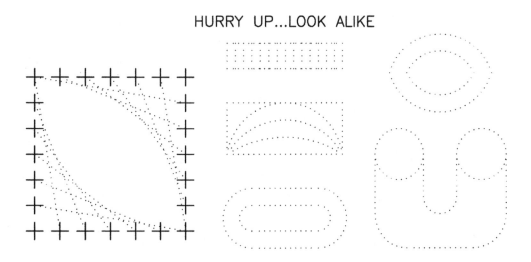

Figure 2.11 Try the exercise again on arcs, circles, and angled lines. Keep your eye on the target and make swift, light pencil strokes. Once the strokes look good, go back and use a softer pencil to make them black.

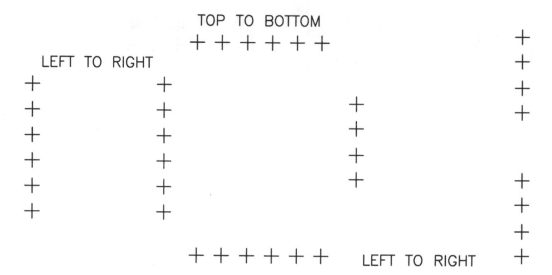

Figure 2.12 Practice sweeping your sketch pencil to the targets without the dotted guidelines. Move the sheet around to find your most efficient sketch angle and pencil stroke.

- Larger circles are easier to create if you begin by drawing very light vertical and horizontal lines through the desired center point. Mark the desired radius on each of these lines **(Figure 2.13)** and then sketch arcs onto each quadrant using a series of connecting dashes. Most right-handed persons do best with arcs that travel upward and to the right from 9 o'clock to 12 o'clock. If this works for you, then turn the sketchpad 90 degrees and repeat the process several times until a full circle is completed. Trace only the elements that are on the desired path with dark lines to make the best circle. You will discover that a good eraser is very useful for this process.

- For very large circles it may be necessary to mark the radius at additional locations in order to control the size of the radius. The edge of a piece of scrap paper can be used to transfer the radius length to your sketch. **(Figure 2.14)** Some people over do this practice and place radius points so close together that they almost form the circle. Don't forget that the intent of sketches is to provide a **quick documentation** of your design, not a finished drawing.

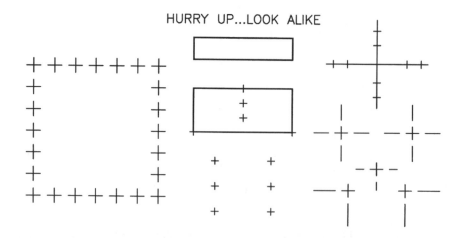

Figure 2.13 Try sketching the arcs, circles, and angled lines again without the dotted lines. Keep your eye on the target and make swift, light strokes. Once the strokes look good, go back and trace them with a softer pencil to make them black.

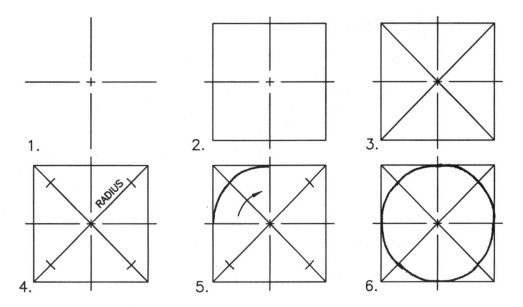

Figure 2.14 The size of circles and curves is easier to control if you first create a construction box. Turn the paper as necessary to use the natural arc motion of your hand.

❒ For ellipses, tangent arcs, and irregular curves, it is best to establish their sizes first with construction boxes. Use these to find the tangent points required for these irregular curves as shown in **Figure 2.15**.

You will need to find out which technique works best for you and then strive to practice until you can create the desired lines and circles quickly and consistently. It may be necessary to start by sketching on grid paper and work gradually to unruled paper. Most engineers continue to use an engineer's tablet with a light grid throughout their careers to create drawings, graphs, and notes.

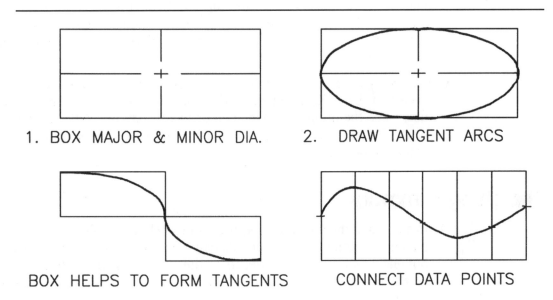

Figure 2.15 Ellipses and arcs are also controlled easier with light construction boxes. Note that the four quadrants of the ellipse always contact the midpoint of each side of the construction box.

Some Tips from the Pros

TIP 1: Place a clean piece of paper over the portion of the sketch paper you are now drawing on, and rest the heel of your hand on it as you work. Oil from your skin will soften the graphite and cause it to smudge easily. The same oils will repel India ink when you need to make a permanent ink drawing over your sketch.

TIP 2: If you are sketching a symmetrical object like the front view of an airplane or parts of the old Texas church, you can complete half of the sketch on a piece of tracing paper and then fold it on the mirror line and trace it to complete the mirrored side. This will insure that both sides are identical and will save you a lot of time.

TIP 3: Trace repetitive features onto a scrap of tracing paper and then trace it as necessary to insure consistency of the features and save time. Using a light table or overhead projector will allow you to see through the bond sketching paper and make tracing much easier.

TIP 4: Place construction lines and rough layouts on one side of some good quality tracing paper, and then turn the paper over and complete the final sketch on the other side. Once the sketch is finished, erase the construction lines on the back quickly and easily without worrying about disturbing the finished work.

TIP 5: Use sketch paper with a very light blue grid. Copy machines can be set to the "lighter image" position and will not duplicate the grid.

TIP 6: To test the darkness of your sketch lines, make a copy on a black and white copy machine. The lighter portions of the copy will show you which parts of the drawing need to be made darker.

TIP 7: If you need to send a copy of your sketch by fax, it's a good idea to make a good black and white copy and run the copy through the fax machine. The copy machine toner usually picks up better on fax machines than pencil lead, and if the paper jams in the fax machine, your original is still intact.

TIP 8: Keep your original sketches in a special file and do not let others have access to them. Keep copies of the originals for distribution and general use. Once the original is damaged or lost it may never be returned to its original quality. By the way, coffee stains on your original often appear as black ink on copies.

Sketching Summary

- In summary, the only tools required for sketching are a pencil, eraser, a sketchpad, patience, and lots of practice. Sketching can be done in any of the artistic media but engineers generally prefer to do their sketches with soft pencils such as an **H** or **HB** on good quality paper. This media allows for clean erasures and easy revisions.

- ❏ Quite often sketches are made on an engineer's sketchpad, which resembles graph paper. Grids on these pads aid drawing straight lines and are available in various grid increments and orientations. Most engineers prefer to work on the backside of the page, which is opposite the side where the grid is printed. This allows their linework and lettering to be in greater contrast with the paper than the dark grid would permit. Pure sketching does not include the use of any drawing instruments, but sometimes, straight edges and circle templates are used to sharpen up a sketch.

- ❏ Occasionally, sketches are traced with black ink so they make better copies and are more suitable for scanning or faxing. When this is done, the pencil construction lines should be erased. If black ink is not going to be used, the light construction lines may be left on the sketch. A good sketch is a valuable communication tool and should be able to stand alone. It should not require any additional explanation from the designer to clarify the drawing or notations. The sketch should also include the date, name of the design, and the name of the designer. This documentation could prove invaluable if a patent challenge or product liability claim should arise.

- ❏ Virtually anything can be broken down into the basic shapes of circles, squares, and triangles. Learning to sketch these shapes quickly and correctly is the beginning of becoming a good sketch artist.

Pictorial Sketches

Pictorial sketches are used to clarify the design or help explain the concept in real time presentations. Early draftsmen had to validate the accuracy of their pictorial drawings with projections of the object's height, width, and depth. This process was extremely labor intensive, slow, and often complicated. For most pictorial drawings created after World War 2, the projection practice was generally replaced with angular drafting where the lines were made parallel with drafting machines or drafting triangles.

There are three categories of pictorial drawings used by engineers: **Axonometric, Oblique, and Perspectives**. **(Figure 2.16)** Each of these categories has three types of pictorials:

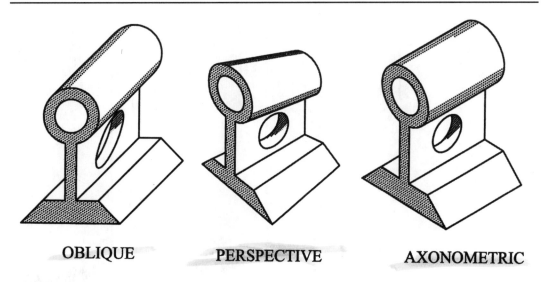

Figure 2.16 Pictorial drawings can be classified into three major categories: Oblique, Perspective, and Axonometric.

- **Axonometric** ... Isometric, Dimetric, and Trimetric
- **Oblique** ... General, Cavalier, and Cabinet
- **Perspective** ... 1 Point, 2 Point, and 3 Point

Axonometric Pictorials

Axonometric projections include the categories of **Isometric**, **Dimetric**, and **Trimetric** pictorials. A characteristic of axonometric drawings is that a cube drawn with any of these projection types will always result in having the lines on each face remain parallel to each other. In other words, any lines that are parallel on the cube will be parallel on the projected views. These projections also result in views that are smaller than the true size of the actual object depending on the rotation and tilt angles used. Circular features have to be plotted as ellipses in all three of these types of pictorials. **Figure 2.17** shows the differences between each of these types of projections. Notice that the "Iso" has only one projection set-up (45° rotation and 35°–16' tilt) ... "Di" has two variations (45° rotation and any tilt angle), and "Tri" has three possibilities (two different rotation angles and any tilt angle). **Figures 2.18**

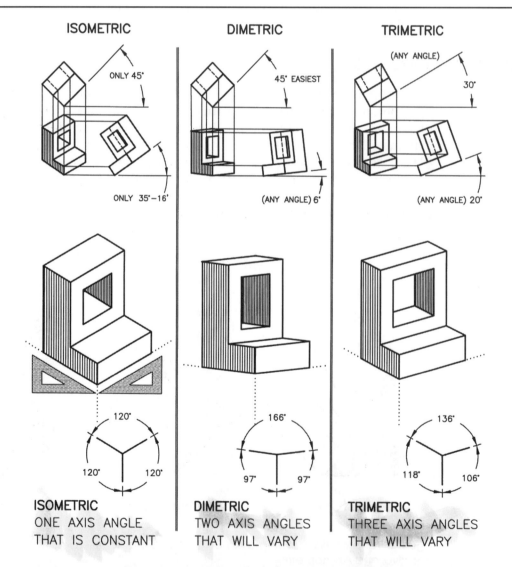

Figure 2.17 Axonometric drawings are classified as Isometric, Dimetric, or Trimetric pictorials. Isometric has become the dominant type used for sketches and finished drawings.

through **Figure 2.20** show the projection techniques required to construct each type of axonometric. It should be easy to see why projection methods have been replaced with angular drawing methods that communicate the design intent and clarify the details necessary for construction just as well in a fraction of the time.

Isometric is the only system of any of the drawing types that maintains uniform scale integrity for each of its surfaces. This means it will not appear distorted or disproportionate in size. Isometrics are probably the most popular type of pictorial used for engineering drawings and sketches. They can be easily constructed because the 30–60° triangle, isometric grid sheets, and isometric ellipse templates are readily available to aid the designer. Because of this popularity, isometric viewpoints are included in most of the computer aided design (CAD) software and parametric solid modeling packages in use today.

Isometric Pictorials

With the obvious advantages of isometric drawings stated above, one might wonder why isometrics are not used exclusively for engineering sketches. One major drawback of using isometrics is that circular features become ellipses. Not only is it more difficult for beginners to draw an ellipse than to draw a circle, but an additional difficulty lies in the orientation of the ellipse on each face of the isometric. Essentially, there are

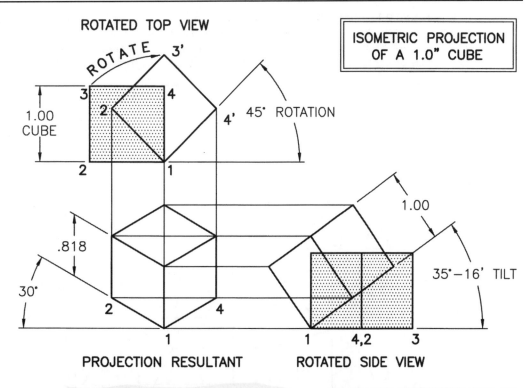

Figure 2.18 The isometric projection of a 1.0" cube only works with one set of angles . . . **45°** rotation of the top view, and a **35°–16'** tilt angle of the rotated side view. With this combination the resultant base angle is 30°. Note that the projected size of the sides of the finished pictorial is 18% smaller than the original cube.

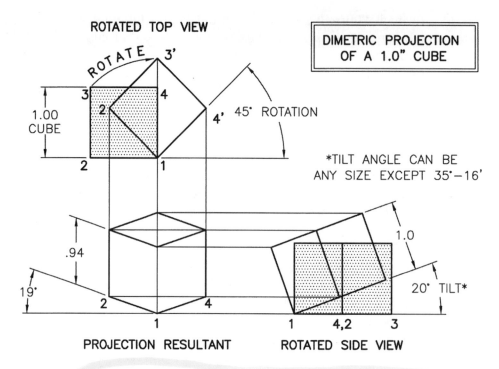

Figure 2.19 The DIMETRIC projection of a 1.0" cube works with two sets of angles ... **45°** rotation of the top view, and **any angle EXCEPT 35°-16'** for a tilt angle of the rotated side view. With the 20° tilt angle shown in this example, the resultant base angle is 19°. Note that the projected size of the sides of the finished pictorial are only 6% smaller than the original cube.

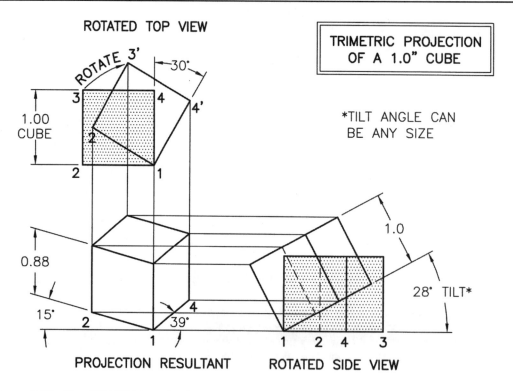

Figure 2.20 The TRIMETRIC projection of a 1.0" cube works with three sets of angles ... **any angle except 45°** rotation of the top view, and **any angle** for the tilt of the rotated side view. With the 30° rotation and 28° tilt angle shown in this example, the resultant base angles are 15° and 39°.

only three positions for ellipses on isometrics: the **top circle**, a **right circle** and a **left circle**. **Figure 2.21** shows an isometric cube with a circular feature on each of these three faces. It illustrates the **easy rule** for locating ellipses on the isometric surface. This rule is to align the major diameter of the ellipse parallel to the longest body diagonal of the cube. In other words, to draw a 2" ellipse, first sketch a 2" isometric box. This box would either have 30° angled sides or vertical sides. The top circle has all 30° sides and the right and left circles use a combination of vertical lines and 30° angles. These boxes become diamond shaped as they are squashed into isometric orientations and the midpoint of each side of the box becomes a tangent point for the

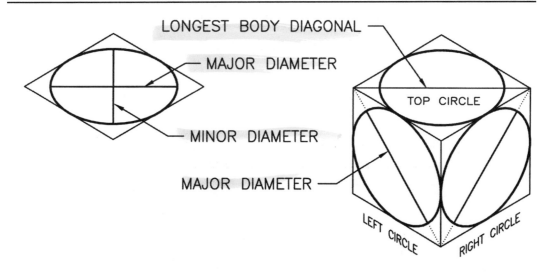

Figure 2.21 Ellipses still touch their quadrants to the midpoint of each side of an isometric pictorial box. The easy rule is to always align the major diameter of the ellipse with the longest body diagonal of the construction box.

ellipse. **Figure 2.22** shows a swivel bracket with holes shown as ellipses on three faces. Some of these holes are so large that you can see through to the backside. The small circles, however, don't allow you to see the backside because of the thickness of the part. The visibility of the backside is determined by two factors: the **ellipse diameter size** and the **thickness** of the part. A large diameter on a thick part might be visible and a small diameter on a thin part could also be visible. The easiest way to establish this is to project the ellipse back at a 30° angle for a right or left side ellipse, or project it vertically for a top circle ellipse. **(Figure 2.23)** Mark off the thickness of the part on this projection and trace the ellipse again. This will determine if the backside is visible. Just think of it like a hole going through a 2" thick board. The top hole would be 2" above the bottom hole. The same concept applies to pictorial projections.

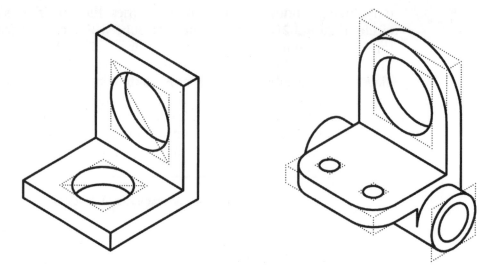

Figure 2.22 Ellipses ONLY APPEAR IN THE TOP, RIGHT, AND LEFT orientation for isometric sketches. Note that the four quadrants of the ellipse always contact the midpoint of each side of the construction box regardless of their orientation.

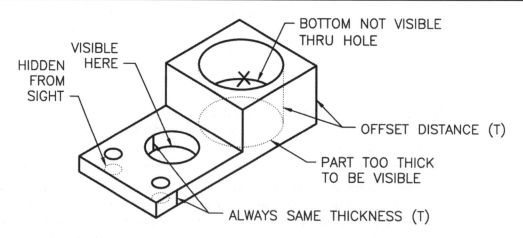

Figure 2.23 The bottom edge of a hole is visible or hidden depending on the part thickness, hole size, and the depth of the hole. Copy the ellipse down or back a distance equal to the thickness (T).

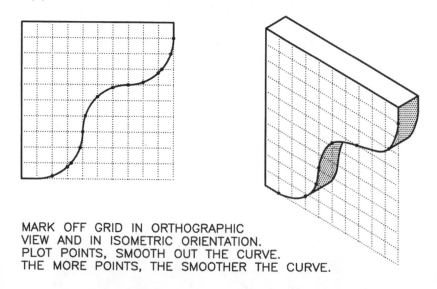

MARK OFF GRID IN ORTHOGRAPHIC
VIEW AND IN ISOMETRIC ORIENTATION.
PLOT POINTS, SMOOTH OUT THE CURVE.
THE MORE POINTS, THE SMOOTHER THE CURVE.

Another awkward concept is drawing angles in isometric. It is possible for an angle to appear larger or smaller than its true size when plotted in isometric drawings. A protractor cannot be used to establish angles in isometrics. Angles have to be plotted by locating the start point, the vertex and the end point. This can be done on grid paper or by using the coordinate method to mark off the dimensions of these three points. (**Figure 2.24**) Another nice feature of using isometric drawings is that they can be viewed from above, below, right or left viewpoints.

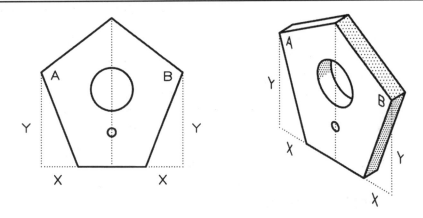

Figure 2.24 Angles and irregular curves must be established with coordinates for most types of pictorials. Note how angles can appear larger (B) or smaller than their actual size (A).

Figure 2.25 illustrates some different positions of isometric drawings. Sometimes, it is necessary to show more than one isometric view of a part in order to show details that are not clear in a frontal view only.

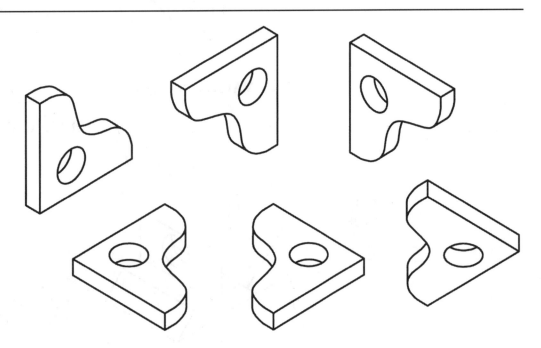

Figure 2.25 Pictorials can be oriented to show the important details in many combinations of the top, bottom, left, or right views.

Oblique Pictorials

Sometimes, oblique pictorials are not as realistic as isometrics, but they make up for it with the speed and ease that they can be drawn. **Figure 2.26** shows some of the contrasts between the two types. Unfortunately, they often become distorted or ugly. The nice thing about obliques is that there are very few rules that govern them. These "non rules" mean that obliques can be drawn at **any base angle** as compared to isometrics that can be drawn at only 30° base angles. The "true" size of obliques can also vary in depth from $1/2$ to full size depending on the style of oblique being used. The three types of oblique drawings are called **cavalier**, **cabinet** and **general**. All three types of obliques are used for engineering sketches. The *cavalier drawing* requires the height, width and depth be true size or scaled at equal proportions to true size. The *cabinet drawing* has the true size for the height and the width, but the depth dimension is reduced by 50%. In other words, if the cube shown in **Figure 2.27** is a 2" cube, in a cabinet orientation, the depth would only be 1". It is not uncommon for a cavalier oblique to look too large because of its depth dimension or for a cabinet oblique to look too small because of its depth dimension. For this reason, a **general** oblique is used. The *general oblique* may vary in size from a $1/2$ size depth all the way

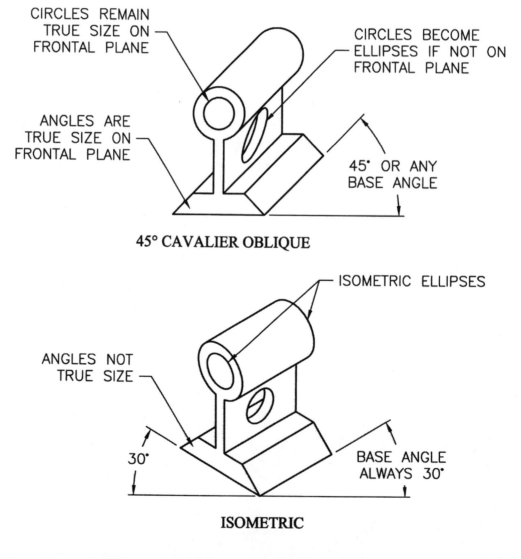

Figure 2.26 Contrasts between Isometric and Oblique pictorials.

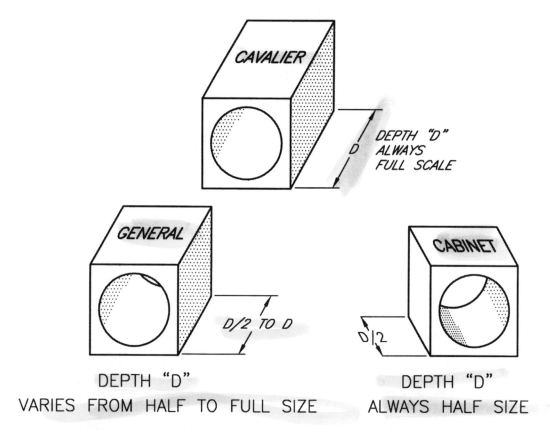

Figure 2.27 Cavalier, General, and Cabinet oblique pictorials are shown at the same receding angle (60°) and identical front views.

up to a full sized depth. In other words, if a general oblique was a 2" cube, its depth could vary from 1" to 2"—it can't be less than $1/2$ depth nor greater than the full depth.

Figure 2.28 shows some objects that look very realistic in oblique projection. Oblique sketching is fast and easy because the front view remains true size and easily accommodates the drawing of circular features and angles. For this reason, it is desirable to always put circular features and angles in a frontal plane of an oblique drawing.

Figure 2.29 shows the proper and improper orientation of an oblique pictorial. Oblique drawings are seen everyday in the form of advertising where the letters of the alphabet are extruded back to become oblique drawings. Sometimes these add thickness and body to the letter, other times it is just a shadow. **Figure 2.30** shows different orientations of oblique pictorials. The engineering sketch should reveal the greatest detail and this will determine which viewpoint should be used to show the oblique

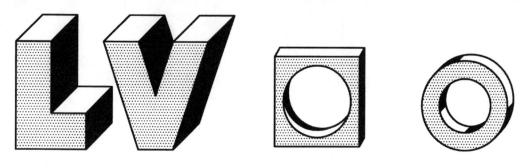

Figure 2.28 Some applications of oblique pictorials, especially thin parts, look pretty good.

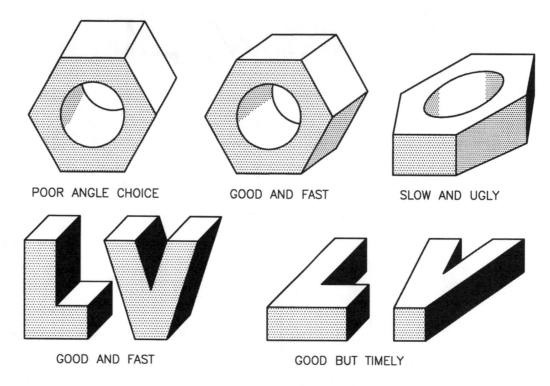

Figure 2.29 Some orientations of oblique pictorials take much longer to draw and may look ugly when finished.

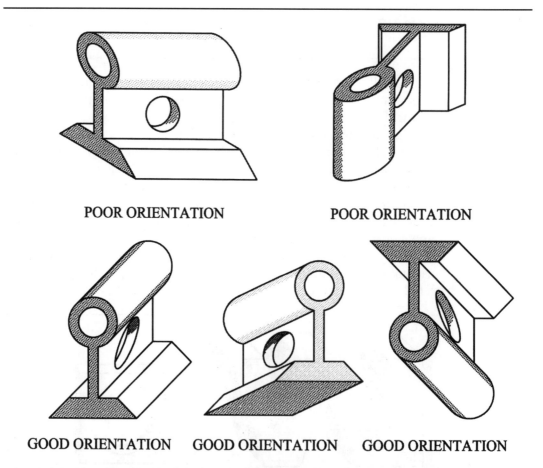

Figure 2.30 For Oblique pictorials, select a viewpoint orientation that shows the most detail and has the greatest number of curves and angles on or parallel to the frontal plane.

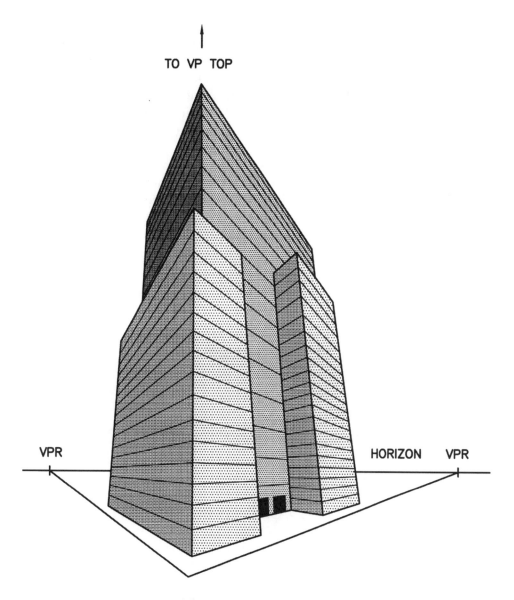

Figure 2.31 Three point perspectives are often used for drawing very tall objects viewed from above or below.

---

pictorial. Occasionally, it might be necessary to have more than one sketch to show all the details. Sometimes you have to make a tradeoff with the rule of having the circles and angles in the front view versus having the most detail. On other occasions, there may be circles and angles in both the front and side views that result in making the face with the greatest number of circles and angles become the front view.

Perspective Pictorials

Perspective comes from the Latin word, prospectus, which means "to look forward." The Renaissance artists created a type of drawing called linear perspective. This type of drawing utilizes vanishing points on the horizon, one on the left, and one on the right for a **two-point perspective**, or one in the middle of the drawing for a **one-point perspective**. When you have a very large structure such as a tall building, there is a three-point perspective, with vanishing points on the left, right and directly above the object as seen in **Figure 2.31**. Perspectives actually come closest to a real photograph as they make distant objects appear smaller just like they appear to the human eye. Although the art of plotting perspectives becomes very complicated, the art of sketching them is very easy. Just do what the artists do, sketch it the way you see it. Essentially, all you have to do is create a horizon line, locate a vanishing point on the right and on the left, and let all the lines that would be parallel in an isometric converge toward these vanishing points as seen in **Figure 2.32**. In a real plotted perspective, there would have to be a side view to project the height, and a rotated top view to project the width. After establishing the ground plane and horizon, lines would be drawn toward the vanishing points on this horizon. **Figure 2.33** shows an illustration of the proper way to establish a perspective vanishing point. However, for most applications of engineering sketches, it is not necessary to be that detailed so estimates are commonly used. Practice drawing lines quickly and lightly to the vanishing point and then darken the visible outlines to define the object. To locate the center of a perspective space,

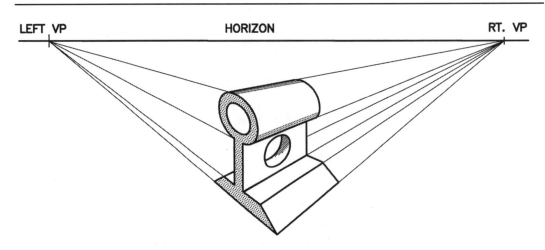

TWO-POINT PERSPECTIVE SKETCHES

Figure 2.32 For two-point perspective sketches, place two vanishing points on the horizon line and extend the edge lines to each point. For sketches, the height and width can be estimated.

the rule of diagonals is used. Diagonals indicate where the perspective center of a wall or a surface would be. **(Figure 2.34)** This is not a true distance because a true distance cannot be measured in a perspective.

Most CAD systems will not manipulate objects, list dimensions, or disclose mass properties while in the perspective viewing mode. Perhaps the most difficult aspect of making engineering sketches in perspective is that of drawing circular features. **Figure 2.35** shows that a perspective circle is not a true ellipse, but rather a truncated ellipse (or egg shape), that has to be plotted. The best way to approximate this truncated ellipse is to draw the diamond box again, but draw it in perspective with each of the four sides having a different length. Obviously, angles cannot be measured in perspective but have to be plotted by locating the start point, the vertex and the end point with coordinates or projections.

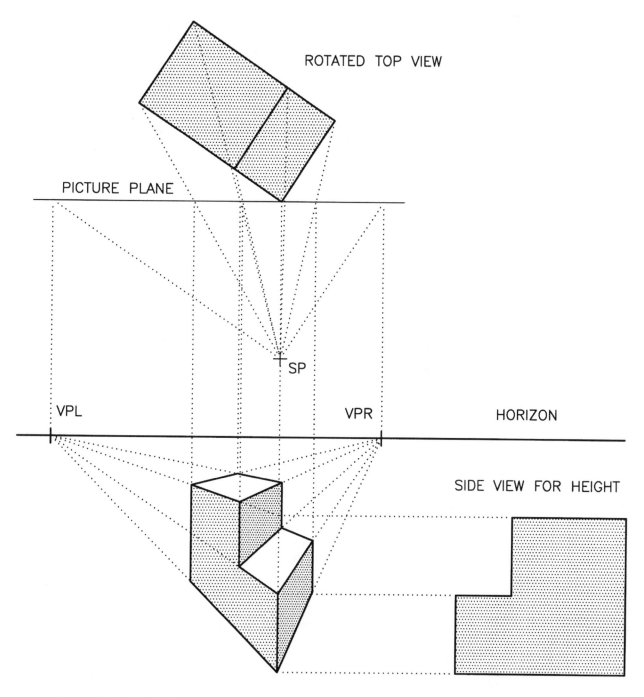

Figure 2.33 Vanishing points are established by locating a station point where the viewer stands. This point can generate a dramatic or subtle perspective depending on its distance from the object.

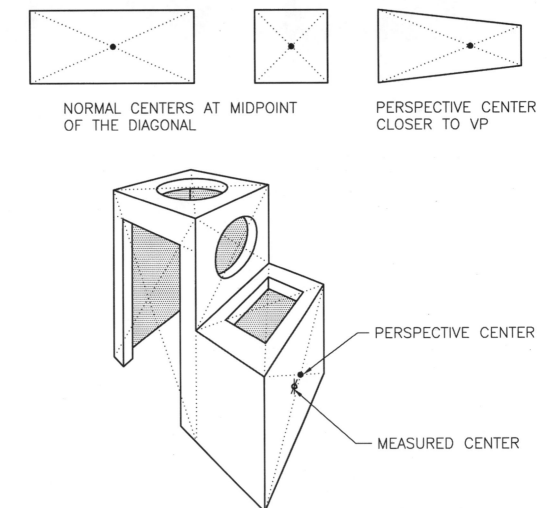

Figure 2.34 Since true distances cannot be measured when sketching in perspective, the centers of surfaces are located with diagonals.

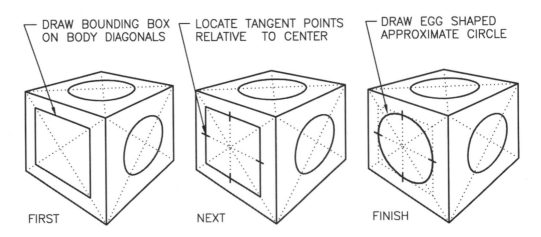

Figure 2.35 Sketching perspective circles involves creating a perspective bounding box to locate the circle's tangent points. The results will be an egg shaped, approximate, perspective circle. Additional projected coordinate points are used if accuracy of the curved path is critical.

Try your hand at sketching the hanger bracket shown below. In addition to the isometric and section views shown, experiment with rendering with shade and shadow.

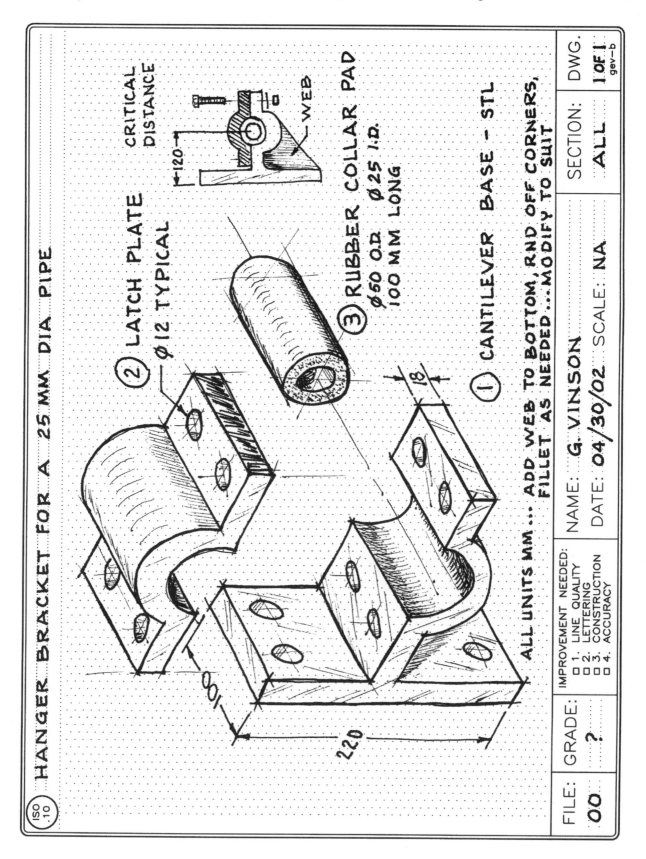

CHAPTER 2—Practice Quiz: Sketching and Pictorials

_____ 1. Light, preliminary layouts are best made with which grade of pencil?
A) H   B) HB   C) F   D) 4H   E) None of these

_____ 2. Sketches need to be made with sharp, dark black lines in order to:
A) Make good copies   B) Fax clearly   C) Show details clearly
D) Communicate design intent   E) All of these

_____ 3. Sketching of circles larger than 1.0" is best done with one quick, continuous clockwise circle.
A) True   B) False

_____ 4. Engineering sketches should include design notes, the originator's name, and the date the sketch was made.
A) True   B) False

_____ 5. Engineering sketches need to be very clear, clean, and concise with all of the construction lines erased.
A) True   B) False

_____ 6. The basic building blocks for sketching are the:
A) Arc, box, cone   B) Cube, prism, sphere   C) Cylinder, sphere, cube
D) Circle, square, triangle   E) None of these

_____ 7. Long, straight, sketched lines are best made with a series of short strokes.
A) True   B) False

_____ 8. When making short lines, a good technique is to keep your eye on the termination point instead of watching the pencil lead.
A) True   B) False

_____ 9. The three types of pictorials commonly used for engineering sketches are:
A) General, common, perspective   B) Isometric, dimetric, trimetric
C) Perspectives, obliques, axonometrics   D) Rendered, outline, screened
E) None of these

_____ 10. The type of pictorial used only for metric drawings is:
A) Axonometric   B) Oblique   C) Perspective   D) Cabinet   E) None of these

_____ 11. A pictorial that has its front face true size and the receding depth is also true size is:
A) Cavalier oblique   B) Cabinet oblique   C) Isometric   D) Dimetric   E) None of these

_____ 12. The most realistic pictorial of a tall building would be made with this type of sketch:
A) General oblique   B) 2 pt. Perspective   C) Isometric   D) Dimetric   E) None of these

_____ 13. The isometric axis angle must always be:
A) 45°   B) 30°   C) 60°   D) 120°   E) None of these

_____ 14. Circles retain their true shape if they are in the front view of which of these pictorials?
A) Cabinet oblique   B) 2 pt. Perspective   C) Isometric   D) Dimetric   E) None of these

_____ 15. Vanishing points are necessary for the construction of this type of pictorial.
A) Cavalier oblique   B) 2 pt. Perspective   C) 3 pt. Trimetric   D) 2 pt. Dimetric
E) None of these

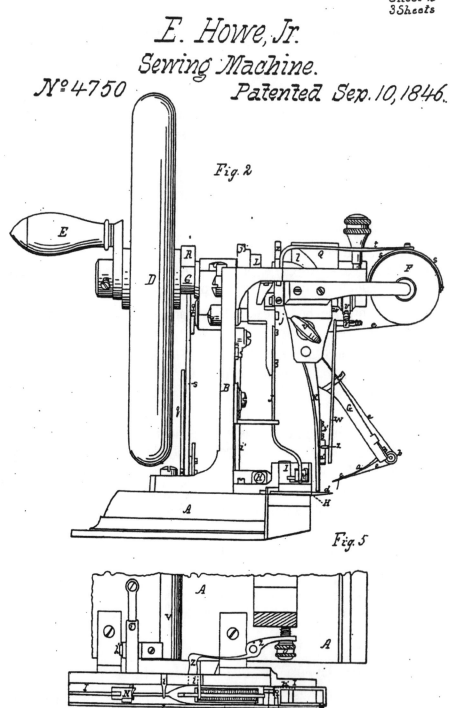

Detailed Sketches

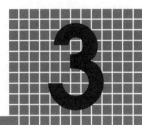

Engineers frequently use flat 2-D views to describe parts of their design. No matter how complex the object or structure may appear, it can usually be broken down into only six different viewing orientations. This is because any object will have a **front** and **back**; a **left** and a **right**; and a **top** and a **bottom**. It is generally not necessary to draw all six views, as there will be some duplication of the shapes and sizes of height, width, and depth. For example, the front and the back views would duplicate the width and the height of the same object, while the right and left views would duplicate their height and depth. Therefore, the rule is to use only as many view as are required to fully describe the object. For a very simple object, sometimes one view is adequate if it is accompanied with a note. For example, a one-view drawing could work with the sketch of a circle if it was accompanied with a note such as "**2.0 sphere**." **(Figure 3.1)** It would be a waste of time to draw a top,

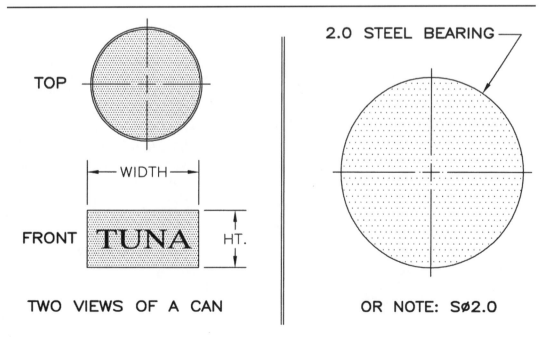

Figure 3.1 Some objects can be fully described with two views or with only one view and a note.

47

front, and side view of a sphere or cube, when a note could communicate the same design information. For more complex objects, the most common combination of views seems to be a three view drawing which shows the **top view**, the **front view**, and the **right side view**, all aligned with each other, and the top always over the front. In order to show these surfaces true size, we pretend to be sketching them on a glass plane placed on top of the object. (**Figure 3.2**) That would be like having a friend put their face against a glass window and tracing the outline of their nose, mouth, eyebrows and ears to capture their image. It is possible that the ancient Egyptians used a variation of this to trace the profile of a person's reflected shadow onto the walls of tombs and temples. If they worked the shadow properly, it could result in a very accurate likeness that was true in size and shape.

Following this logic, if we needed to describe all six of the object's standard faces (top, bottom, front, rear, left, and right), it would be necessary to use six different glass panes to surround the object. The results would form a glass cube with sides mutually perpendicular with the object resting inside. It might be easier to visualize this if you imagine a model car placed inside of an empty aquarium as in **Figure 3.3**. If you align the sides of the model car, parallel to the sides of the glass, you could work your way around the aquarium and trace the details of each side of the model on the outside of the glass. Admittedly very awkward, this would no less, capture all six of the standard views of the model car, describing everything from grille to tail lights. Although this would be one way to describe objects, it would certainly not be very efficient to tote and store large glass cubes around the office or shop. In order to lay all of the views out flat, the glass box would have to be unfolded as shown in **Figure 3.4**. When the views are properly aligned, you will notice that there is some duplication of the shapes and sizes. In **Figure 3.4** for example, the **height (H)** of the toy SUV is shown in the front, right, back, and the left side views. The **width (W)** can be seen in

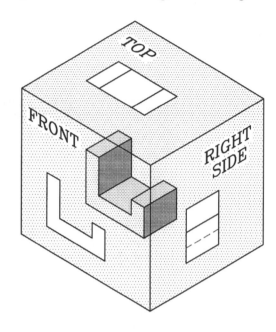

Figure 3.2 Perpendicular projections onto these "glass" reference planes is called orthographic projection. Any two adjacent views will provide the design information needed for height, width, and depth.

Figure 3.3 Projecting the outline of the model car onto the glass planes would give the primary orthographic (flat) views... **TOP, FRONT, AND SIDE**. In this example, a BACK view would be helpful to show the spare tire details of the toy car.

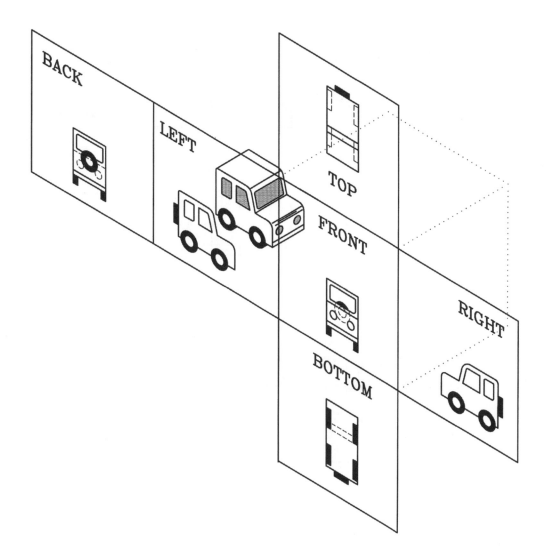

Figure 3.4 By unfolding the glass box we can see all six of the principle views. The dimensions of height, width, and depth are all duplicated.

---

the front, top, bottom, and back views. The **depth (D)** can be seen in the top, bottom, left, and right side views. **(Figure 3.5)**

In trying to remember which dimensions go with each view, the easy trick is to imagine that you are about to shake hands with somebody standing in **front** of you. You can instantly tell how tall (H) and ugh . . . how wide (W) they are. If they should turn to the side, you still see their same height, and now you also see the distance from front to rear we call depth (D). You would have to be looking down at them from a balcony in order to see their top view which would provide width (W) and depth (D) sizes. So that's all there is to it . . .

- **Front and Back** views show height and width **(H, W)**
- **Top and Bottom** views show width and depth **(W, D)**
- **Left and Right** views show height and depth **(H, D)**

The designer has the responsibility to select the **most descriptive views** used to detail the design and the orientation that best describes them. **Figure 3.6** shows a properly oriented set of views in contrast to **Figure 3.7** where the correct descriptive

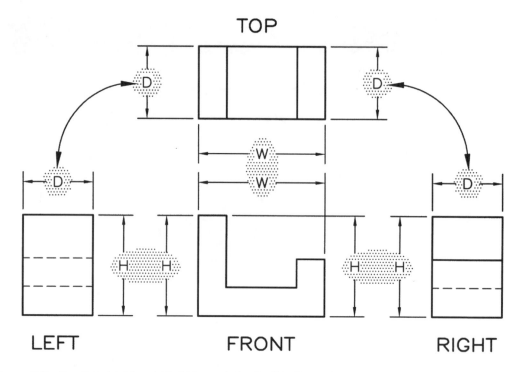

Figure 3.5 The height (H), width (W), and depth (D) dimensions are duplicated here with unnecessary views. Actually, this object could be fully described with only two views...which two would you keep?

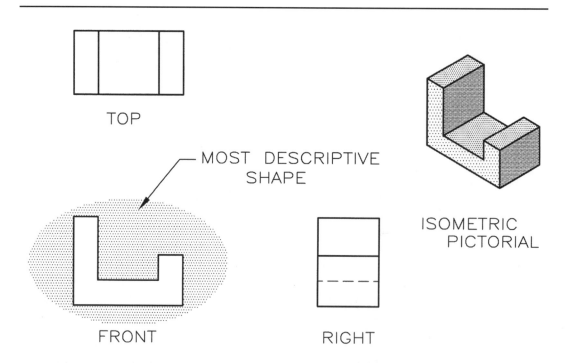

Figure 3.6 Proper orientation of the part places the most descriptive shape as the front view and has a minimum of hidden lines.

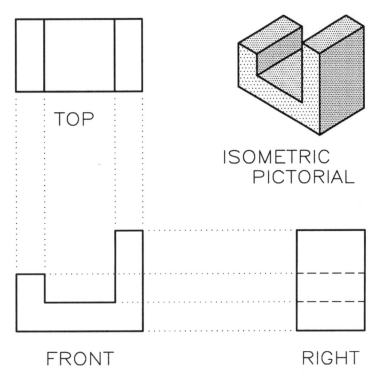

Figure 3.7 Poor orientation of front view results in additional hidden lines and obscured areas in the right side view.

shape, but in the wrong orientation, is shown. For mechanical parts we need to utilize only the views that would be necessary to clearly describe the **shape** and **size** of the object and provide the information required for its construction.

Using Different Linetypes

1. **HIDDEN LINES** When a feature of an object is blocked from sight because it is on the backside of a principal view, then it is shown with dashed lines called **hidden lines**. These lines must be very black and they are about half as thick as the visible outlines that show the object. Hidden lines consist of a series of uniformly alternating dashes and gaps. **(Figure 3.8)** If hidden lines happen to coincide with visible or centerlines, only the boldest line should be shown. **(Figure 3.9)** The origin and termination of hidden lines should appear as shown in **Figure 3.10**. When hidden lines cross over other hidden lines or originate at other hidden lines, they should pass through the gaps as shown in **Figure 3.11**.

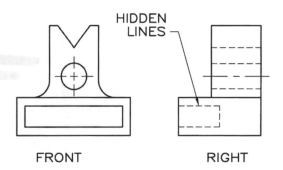

Figure 3.8 Hidden lines shown above are describing the obscured shapes of three different features. Without the side view and hidden lines you would not know that the rectangular slot only cuts half way through the part.

2. **VISIBLE LINES** **Visible outlines** are sometimes referred to as **object lines** since they show the outline of the object. They are probably the most important linetype

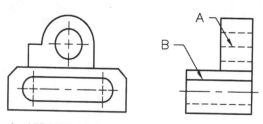

A. HIDDEN LINE REPLACES CENTERLINE.
B. VISIBLE LINE REPLACES HIDDEN LINE.

Figure 3.9 When two lines coincide, show only the boldest one. Visible lines cover hidden lines and hidden lines cover centerlines.

since they show the shape of the object and they also are the lines that show up best when the print fades or the fax is very weak. We draw visible lines with a slightly dull H or 2H pencil in order to make them wider than hidden and centerlines. You may have to trace over the lines several times to make them black. Visible lines appear in nature when there is a sharp bend or break in a surface. These can be sharp like a rock ledge sticking out of the ground or gradual like a gently curving hillside against the sky. In engineering drawing, we follow the same concept by outlining objects with visible lines around their silhouette and showing the bend lines on surfaces. The orthographic drawing of the old coupe in **Figure 3.12** is made up exclusively of visible lines except for the shade area under the fenders where dots are used. In addition to the silhouette being defined, notice that cracks around the door, edges of the fenders, and the louvers in the hood, all made with visible lines, add realism to the drawing since no shade or rendering is used.

3. **CENTERLINES** Another line, which is necessary for orthographic views, is the **centerline**. It shows the location of the **center** of arcs and circles, and can be used for dimensioning the center point of a hole. **(Figure 3.13)** It will also allow us to determine if a rectangular appearing object is truly a rectangle or if it is a cylinder. It is very difficult to accurately locate where to drill a hole from measurements to the outside edge of a drill. Before holes are drilled, the machinist marks the centers of

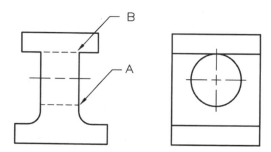

Figure 3.10 Hidden lines should start and end with the dash against a solid line (A). However, if it is aligned at the end of a visible line as a continuation, it starts with a gap (B).

holes on the metal (or enters coordinates for a CNC machine) based on dimensions from the centerlines on a drawing. Once the centers are marked, the machinist uses a center punch and hammer to make dimples (called center marks) in the

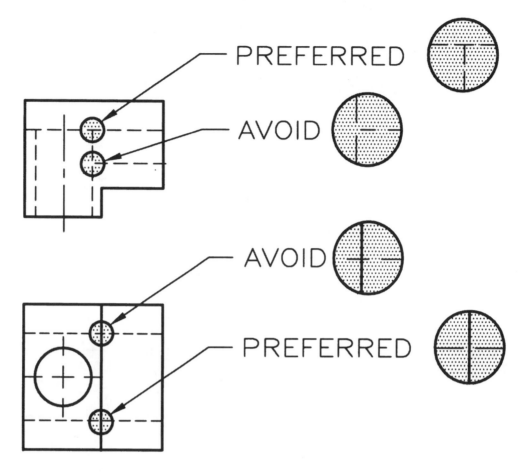

Figure 3.11 Hidden lines should cross visible and other hidden lines as shown above.

Figure 3.12 This CAD drawing of an old car (circa 1930) is made entirely of visible lines except for the shade under the fenders. In addition to showing the silhouette, they are used for details like the door outline and louvers on the hood.

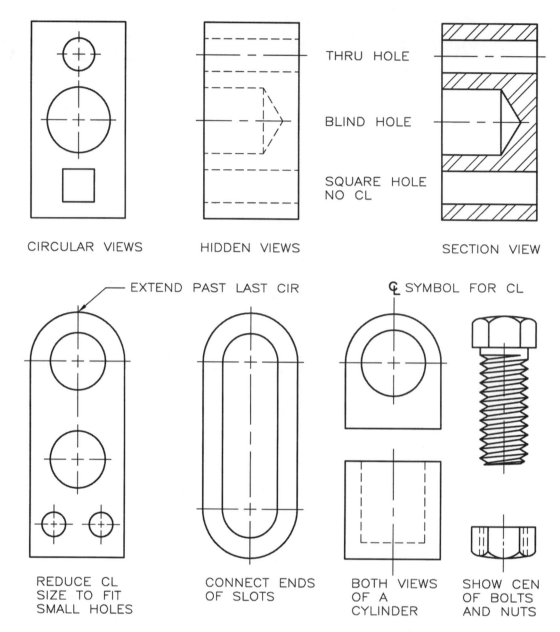

Figure 3.13 Centerlines (CL) should be attached to both the circular and rectangular view of holes. Apply them to visible and hidden features as illustrated in the views above. Reduce the linetype scale to fit with small holes and arcs.

metal at the intersection of the crossing lines (the plus sign in the middle of two intersecting centerlines). This helps to stabilize the drill and insures that it starts at the right location. In higher tech applications, the part to be drilled is clamped down or held with a fixture. Again, the setup and alignment of these is also based on knowing where the center of the hole is located to align with the fixture . . . all based on centerlines.

All holes on engineering drawings, whether visible or hidden, should have centerlines making a "plus mark" in their circular view and another centerline showing the cylindrical axis of the hole. The centerlines should always extend past the last circular edge approximately one text height and should also extend along the hole axis, past the end of the hole in its rectangular view by about the same distance. **(Figure 3.14)**

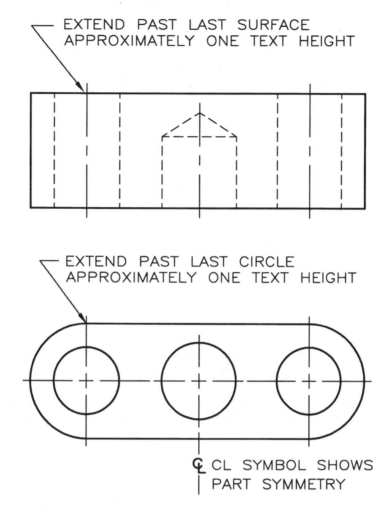

Figure 3.14 Centerlines should always extend past the circular edge and the flat edge approximately one text height.

Getting Started

When sketching orthographic objects, it is best to outline the height, the width and the depth in the three views you wish to show. Box in lightly with a 4H pencil and the construction lines won't have to be erased. Once the object has been defined, trace over the object with a heavy 2H or H pencil to make good black outlines. When deciding which combination of views to use, first establish the front view as the view with the most descriptive shape, fewest hidden lines, and the most detail. Once the front view is established, project the height into the profile view (either the right or left side) and the width up into the top view. **(Figure 3.15)** Again, the most common combination of views are the top, front and right side views. When the views are properly aligned, they must have the same ground plane and the same top elevation. Think about a person standing in front of you. If the person is 6' tall and you are looking at their front, how tall would they be if they turned and you are looking to their right side? They would still be 6' tall. These objects are the same way. They have to be on the same ground plane and the height remains constant between the front and the profile views as shown in **Figure 3.16**.

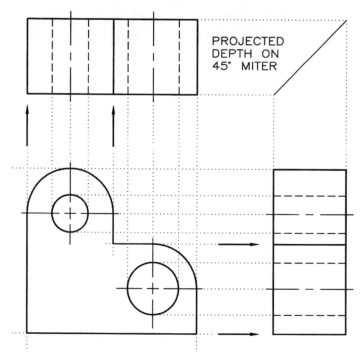

Figure 3.15 Use projections from the front view to help establish the height and width of the side and top views. A 45° miter line can be used to transfer the depth.

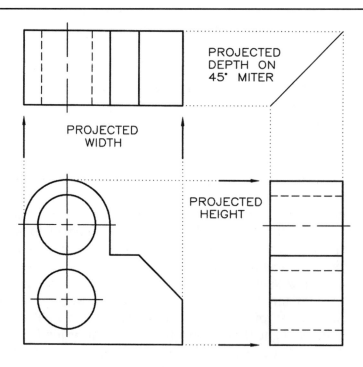

Figure 3.16 Views must be aligned and on the same ground plane in order for the principles of orthographic projections to work. Can you explain why a hidden line and a centerline seem to be missing?

Proper Definitions

Figure 3.17 shows the proper alignment of a picture of a birdhouse for both third angle projection and first angle projection. The obvious front view will be the same in both systems, with the house in its normal resting position, roof on top, and the circular details of the "door" of the birdhouse in the front view. The back view of the birdhouse will show the circular door as a hidden line. In the United States, we use third angle projection that seems natural to put the right view on the right and the left on the left. With first angle, this logic is reversed showing the left view on the right, the right view on the left and the top on the bottom. Obviously, we prefer the third angle best, but most of Europe, Australia and New Zealand use first angle. Because of this, we need to recognize the symbols as shown in **Figure 3.17** and understand their view arrangements.

Reviewing the Basics

1. Choose the most descriptive view as the front view.
2. Align views both vertically and horizontally.
3. Always place the top view over the front view.
4. Show the object at rest in its normal sitting position.
5. Make all views the same scale.
6. Do not include unnecessary views.
7. When lines in a view coincide, keep the boldest and discard the lesser line.
8. Space views uniformly on the page to avoid crowding.

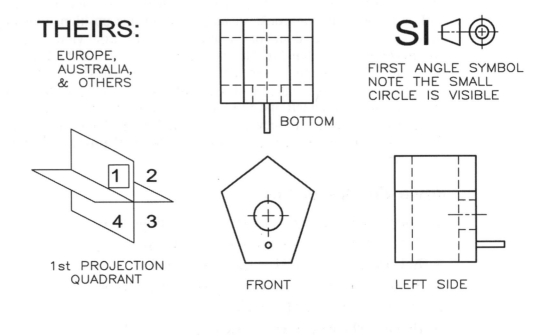

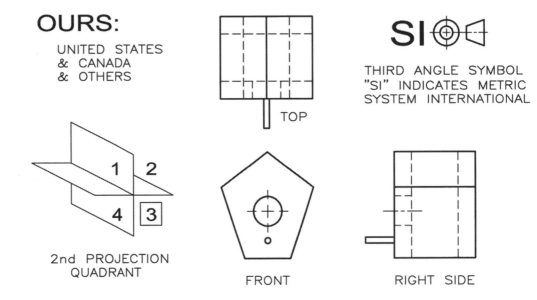

Figure 3.17 The front view should appear at rest in its normal position. The third angle projection system used in the U.S.A. seems to be more natural for the orientation of the other views. The birdhouse in this illustration is made of wooden ends wrapped with very thin sheet metal.

Orthographic Drawing Summary

To summarize orthographic drawings, let's recall that there are 6 possible views of any object: the top and bottom, left and right and a front and back. These are called the principal views but they are aligned with 3 principal planes. These planes are the **horizontal plane**, the **frontal plane** and the **profile plane**. The horizontal plane will be used to show the top and the bottom views of the object. The frontal plane will show the front and the backside of the object (like two sides of the aquarium that are parallel with each other). The profile plane will be the left and the right view. When an object is not parallel with one of the principal planes, it may be necessary to use an extra view known as an auxiliary view. Auxiliary views in **Figure 3.18** are aligned parallel with the inclined surface. These are covered in greater detail in Chapter 6. Occasionally, it may be necessary to remove a portion of an outside surface to show some complicated interior details. For this, section views are used as shown in **Figure 3.19**. Section views are also covered in Chapter 6.

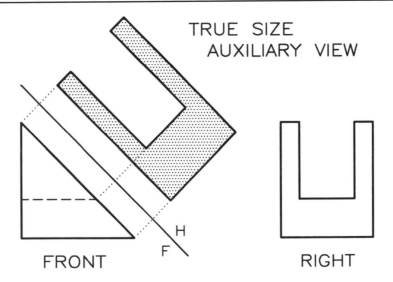

Figure 3.18 Auxiliary (meaning extra) views are used to show true shapes and sizes that appear foreshortened in the principle views.

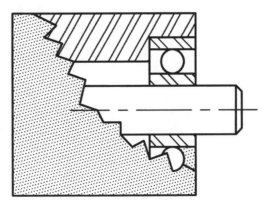

Figure 3.19 Section views are used to reveal interior details that might seem confusing with hidden lines.

CHAPTER 3—Practice Quiz: Multi-View Drawings

____ 1. How many principle views can be used to describe most objects?
   A) 2   B) 4   C) 6   D) 8   E) None of these

____ 2. The reference plane system used to describe orthographic projection principles might best be compared to a glass ____ .
   A) Cylinder   B) Box   C) Sphere   D) Wedge   E) None of these

____ 3. The number of views required to completely describe any object is ____ .
   A) 3   B) 4   C) 6   D) Depends on the object   E) none of these

____ 4. Features on the backsides of views that cannot be seen are drawn with ____ lines.
   A) Invisible   B) Dotted   C) Hidden   D) Ghost   E) None of these

____ 5. In orthographic projection, the top view should always be aligned over the ____ .
   A) Front view   B) Right side view   C) Profile view   D) Bottom view
   E) None of these

____ 6. The dimensions of width and height could be found in the ____ view of an orthographic drawing.
   A) Top   B) Right side   C) Front   D) Front or back   E) None of these

____ 7. Lines used to indicate the axis of a cylindrical hole are called ____ .
   A) Axis lines   B) Middle lines   C) Drill path lines   D) Circle guide lines
   E) None of these

____ 8. When one type of line is projected to lay on top of another line (coinciding), which one do you show?
   A) The Boldest   B) The thinnest   C) The blackest   D) Both of them   E) Doesn't matter

____ 9. The visible outline of objects in orthographic layouts should be drawn as ____ .
   A) Dotted and light   B) Bold and black   C) Light dashes   D) Thin and black
   E) None of these

____ 10. The principle projection planes used in orthographic layouts are ____ .
   A) Mutually perpendicular   B) Always horizontal   C) Always vertical
   D) Depends on the object   E) None of these

____ 11. In the interest of completely describing the object, it is common practice to show it in all of the principle views.
   A) True   B) False

____ 12. It is possible to fully describe some objects with one view and a note.
   A) True   B) False

____ 13. In order to conserve space in third angle projection, it is acceptable to place the left view to the right side of the front view as long as it is labeled "left side view."
   A) True   B) False

____ 14. The view selected to be the front view is chosen because it has ____ .
   A) The fewest holes   B) The most descriptive shape   C) The fewest dimensions
   D) The greatest height   E) None of these

No. 821,393. PATENTED MAY 22, 1906.
O. & W. WRIGHT.
FLYING MACHINE.
APPLICATION FILED MAR. 23, 1903.

3 SHEETS—SHEET 2.

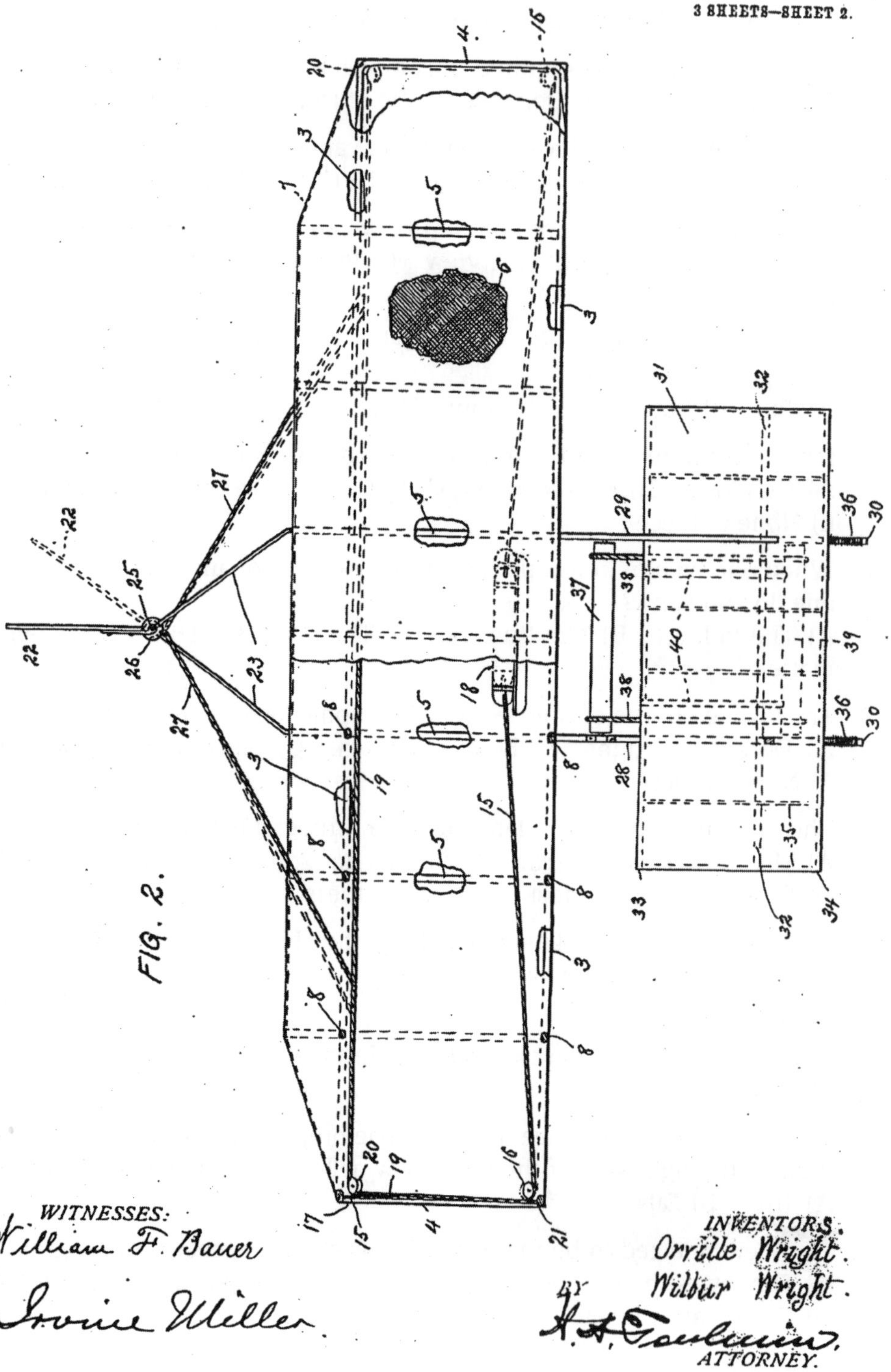

FIG. 2.

WITNESSES:
William F. Bauer
Irvine Miller

INVENTORS
Orville Wright
Wilbur Wright
BY
H. A. Toulmin
ATTORNEY

Specifications and Dimensioning

Dimensions are required on engineering sketches and finished drawings to show the size and location of all features. Before a part can be manufactured, it must be completely dimensioned and include complete specifications for the materials. If dimensions are not correct or some are omitted, a faulty part or mechanism may result which could lead to a catastrophic failure. It is the engineer's responsibility to be certain each designed part is completely specified to reflect the design intent. "First, get it right on paper boys, and then it will always work!" (Early advice from Mrs. Susan Catherine Wright to sons Orville and Wilbur.)

Figure 4.1 shows the size and location of features of the part by using the combination of text, extension and dimension lines, and arrowheads where required.

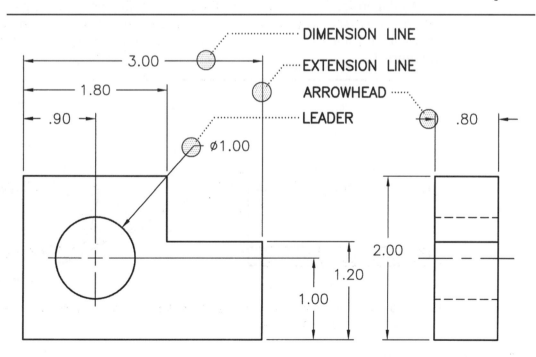

Figure 4.1 A fully dimensioned part is described with the use of **extension lines, dimension lines, leaders, text,** and **arrowheads**.

To print dimensions by hand or with CAD, the standard text height is ⅛" (3mm for metric drawings). Many of these rules are based on text height ratios. The first line of text for dimensions should be 3 text heights away from the edge of the part. Additional lines should be 2 text heights away from the previous lines as shown in **Figure 4.2**. Extension lines pull the dimensions away from the part and they are offset ½ of the text height from the edge of the part. The other end of the extension line goes past the last dimension one text height. Consistent lettering height is one key to having consistently attractive dimensions.

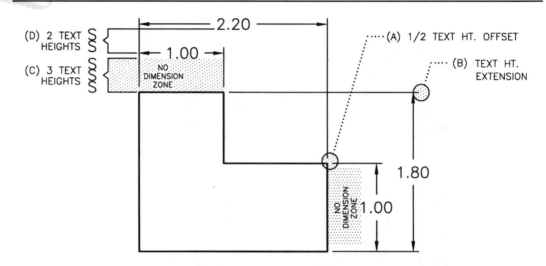

Figure 4.2 Extension lines should be offset from the part by **one half** of the text height (A) and should extend the one text height beyond the dimension line (B). The first row of dimensions must be a minimum of **3 text heights** away from the part (C) and any dimensions that are beyond it a minimum of **2 text heights** apart (D).

All dimensions should be placed on the drawing using very light guidelines and all arrowheads should be made in accordance with the ANSI standard for arrowheads. This standard states that an arrowhead is three times longer than it is wide as shown in **Figure 4.3**.

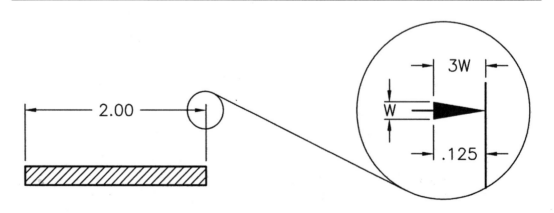

Figure 4.3 An arrowhead must be three times longer than it is wide. It is one text height long (.125") so it should be only .04 wide.

This means that if an arrowhead is .125" long, it should only be about .04" wide at the widest point as shown in **Figure 4.4**. Now, because of the scaling ease of computer graphics, the size has to be increased or decreased to match the printout scale. It is no longer a constant but a ratio of the text height. The current rule is to make arrowheads the same length as the text height. This may seem like a lot of emphasis on a trivial symbol, but if the arrowheads cannot be clearly seen, the dimension can be misunderstood or read to the wrong reference. Bold, black, and uniform arrowheads are one of the keys to clear, concise communication of the size and location of the design intent.

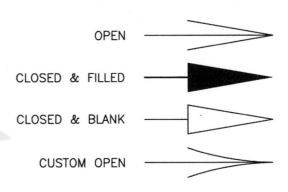

Figure 4.4 An arrowhead may be open or closed, filled or unfilled. Arrowheads made with pencil are generally open and CAD made are generally closed and filled.

Rules for Dimensioning

The rules for placement and location of dimensions on engineering drawings are generally uniform across most industries and are very strict. Some of the rules deal with the size and placement of dimensions and are easy to learn. Industrial rules may vary slightly depending on the products manufactured (for example, Steel fabrication vs. Electronics) but the ANSI general rules are preferred for government and international contracts. The ANSI and some additional rules of good practice are:

1. Avoid placing any dimensions on the part (inside of the view) unless there is no other option. **(Figure 4.5)**

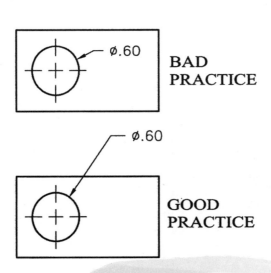

Figure 4.5 Avoid placing dimensions on the part.

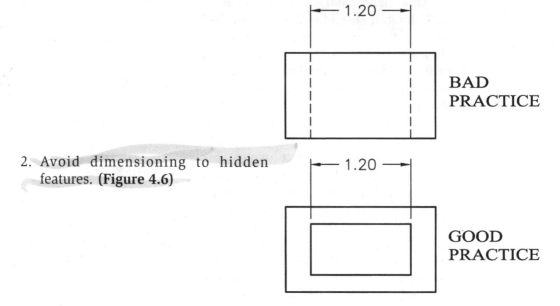

2. Avoid dimensioning to hidden features. **(Figure 4.6)**

Figure 4.6 Avoid dimensioning to hidden features.

3. Always place the dimension where the characteristic shape is shown in the **most descriptive view**. This means *don't* place a dimension on object lines making a "T joint" as shown in **Figure 4.7**.

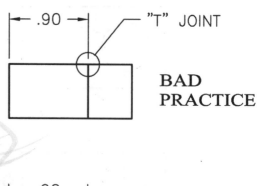

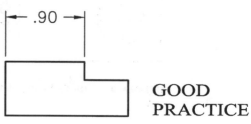

MOST DESCRIPTIVE VIEW

Figure 4.7 Place dimensions on the **most descriptive view**. Avoid placing them at "T" joints.

4. Always dimension holes in their circular view by stating the diameter of drilled holes. Specify the hole depth or special features such as countersinking with a note following the dimension. (Ø.750 x 1.00 DEEP) **(Figure 4.8)**

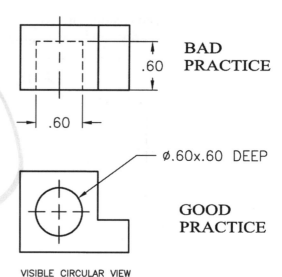

Figure 4.8 Dimension holes with their diameters in a circular view.

5. Dimension rounded corners and arc features as radii where they appear in their rounded views. **(Figure 4.9)**

6. If the same value is repeated many times, then use a general note for the features ("ALL FILLETS AND ROUNDS ARE .125R").

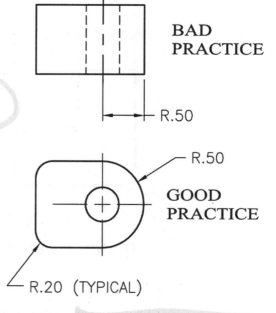

Figure 4.9 Dimension rounded corners and arcs as radii in their circular views.

7. Dimension cylindrical objects as diameters in their rectangular view. **(Figure 4.10)** If only one view is shown, be sure to indicate the value is a diameter. ("Ø2.550")

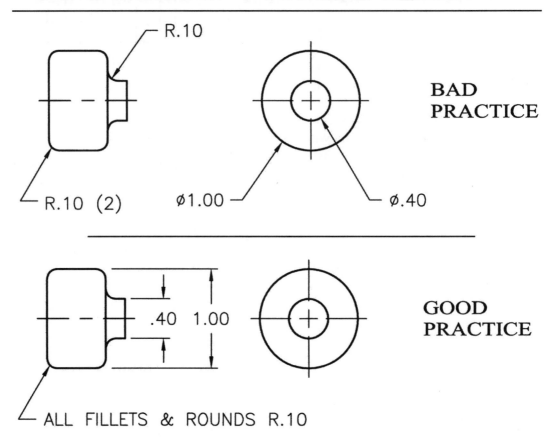

Figure 4.10 Dimension CYLINDERS in their rectangular view with a diameter. Use a note for multiple features.

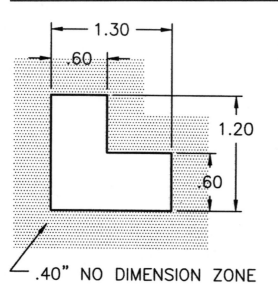

8. Always place the first row of dimensions a minimum distance of 3 text heights (³⁄₈" or about 10mm) away from the edge of the part. Additional stacks of dimensions can be a minimum of two text heights (¹⁄₄" or about 6 mm) away from each other. **(Figure 4.11)**

Figure 4.11 Dimensions should never be closer to the part than 3 text heights (about .40"). Separate the following rows of values by 2 text heights (about .25").

Specifications and Dimensioning 69

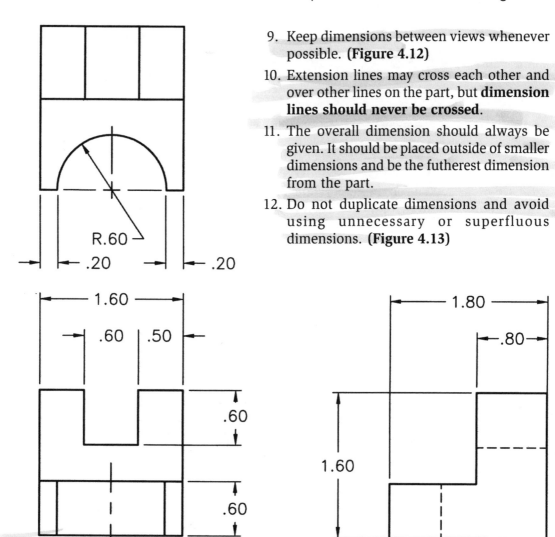

9. Keep dimensions between views whenever possible. **(Figure 4.12)**
10. Extension lines may cross each other and over other lines on the part, but **dimension lines should never be crossed**.
11. The overall dimension should always be given. It should be placed outside of smaller dimensions and be the futherest dimension from the part.
12. Do not duplicate dimensions and avoid using unnecessary or superfluous dimensions. **(Figure 4.13)**

Figure 4.12 Dimensions should be placed between views whenever possible. If space is available, pull dimensions farther away from the views than the minimum distance allowed for a cleaner layout.

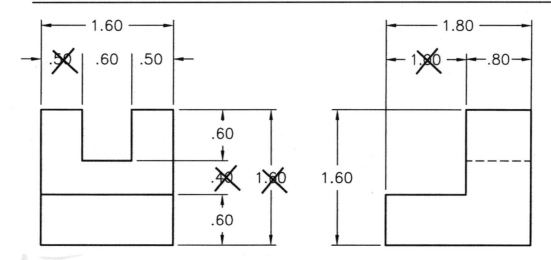

Figure 4.13 Dimensions should not be duplicated. Extra dimensions should be omitted as they create problems with tolerance build-up.

13. When all of the dimensions are expressed as inches, do not use inch mark (") or the abbreviation for inches (in.).
14. For drawings dimensioned in inches, values less than one inch should not be preceded with a zero (.250).
15. For metric dimensions less than 1 mm, place a zero in front of the decimal point (0.255).
16. For metric drawings, omit the use of the millimeter ("mm") notation following the numeral, as millimeters are the default units.
17. The origin for baseline or ordinate dimensions used as a datum should be extended from a finished edge ("V") of the part. **(Figure 4.14)**

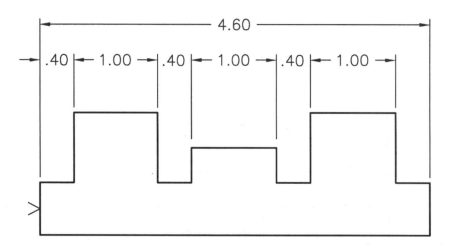

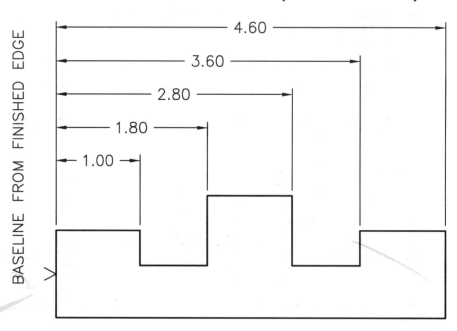

Figure 4.14 Baseline or ordinate dimensions should be referenced to a finished edge. Finished edges are noted with finish marks (V or V').

18. Conserve space and time by using abbreviations and standardized symbols whenever possible. **(Figure 4.15)**

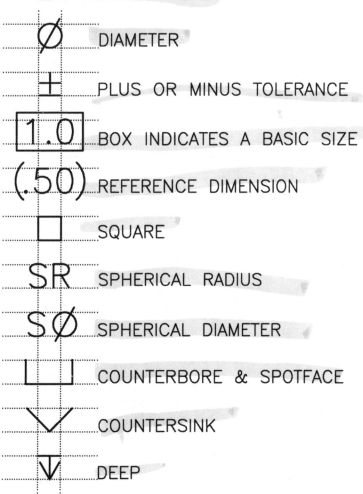

Figure 4.15 Symbols such as these should be used whenever possible in order to save time and space.

19. Reference dimensions should be placed in parentheses or should include the abbreviation "REF." Basic sizes (to be toleranced) should be placed inside of a rectangular box. **(Figure 4.16)**

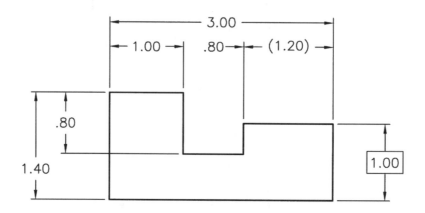

Figure 4.16 Extra dimensions provided for reference only are placed in parenthesis. Basic sizes of features to be toleranced are placed inside rectangular boxes.

20. Extend leaders from the first or last word in a note. Point them toward the center of the circular features that they are specifying. **(Figure 4.17)**

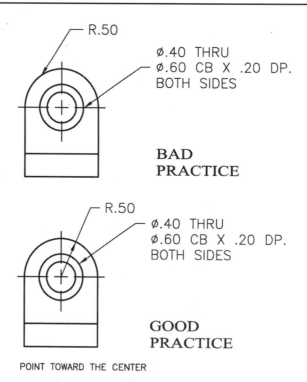

Figure 4.17 Leaders should be connected to the first or last word of the note. They should point toward the center of circular features.

21. Place dimensions among the various views to avoid crowding. Stagger horizontal dimensions to avoid contact or crowding of the values. **(Figure 4.18)**

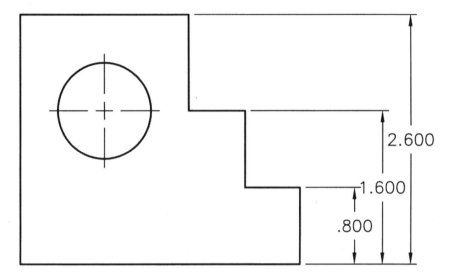

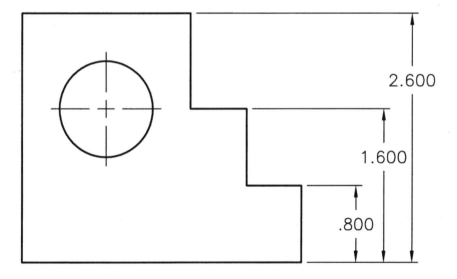

Figure 4.18 Stagger dimensions where necessary to prevent crowding.

Placement of Dimensions

Dimensions may be placed on a drawing in one of two methods, **aligned** and **unidirectional**. The **aligned** method allows the dimensions to be read from the bottom or the right side of the page only. In the **unidirectional** method, the directions are all read from the bottom of the page. **(Figure 4.19)**

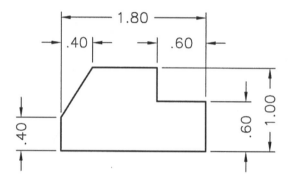

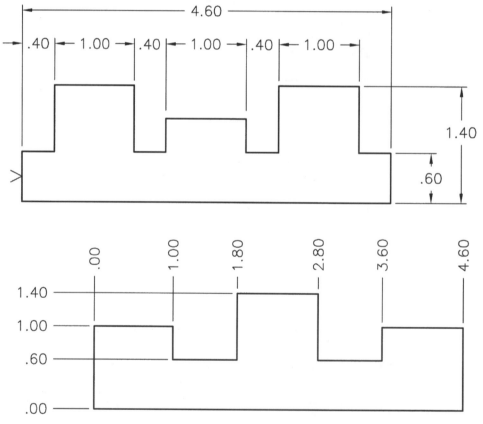

Figure 4.19 Ordinate dimensioning is especially well suited for interfacing with computer numeric control (CNC) machines which can derive their dimensions directly off of the drawing file.

Of the two methods used by industry, the unidirectional method is the most popular and it appears as a default in most CAD programs. In either method, the text should be all caps, black and bold, and the arrowheads should also be black. The dimension and extension lines should be drawn thin like centerlines. In other words, the width of dimension lines, extension lines, and leaders must be about $^1/_2$ of the width of the regular visible outlines. Occasionally, centerlines become extension lines to locate the position of holes on a part. Also, dimensions can be placed in a chain, in an ordinate fashion, or they can be placed relative to a common baseline (datum) as shown in **Figure 4.20**. While chain dimensions are a very efficient use of space, they can contribute to tolerance buildup and can result in faulty parts. Where precision is required, it is preferred to use datum-referenced dimensions or ordinate dimensions as shown. Datum dimensions should always be referenced to a finished edge of the part.

Figure 4.20 Baseline dimensions are recommended when precision is required as tolerance buildup can be better managed.

Table Driven Dimensions

Some industries list their dimensions in tabular form on spreadsheets such as Microsoft Excel®. When properly linked to a parametric solid model, the sizes listed on the spreadsheet can be changed or suppressed and the changes will appear on all drawings that are linked to the spreadsheet. This is a particularly useful time saver when various sizes and configurations of the same part are required.

Dimensioning Standard Parts

This section will show how to dimension fillets, rounds, necks, grooves, and special hole types. The special hole types include countersunk, counterdrilled, counterbored, tapped, blind holes and through holes. Examples of dimensioning these standard features are shown in **Figures 4.21–4.33**.

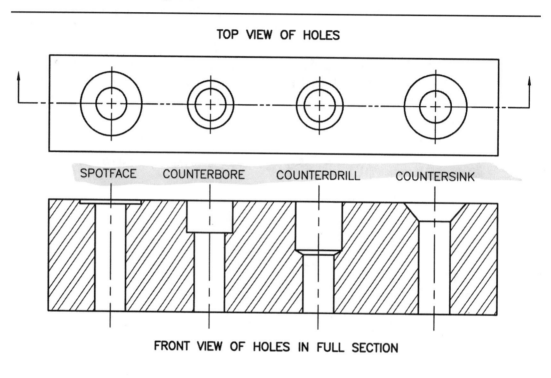

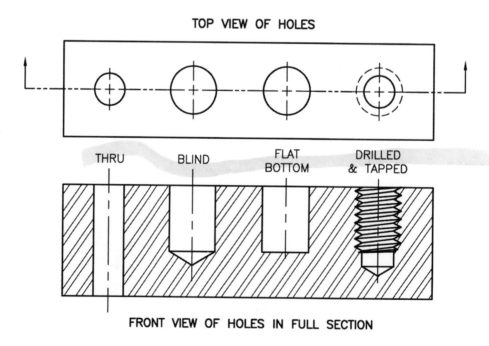

Figure 4.21 Types of holes commonly used for manufacturing.

HOLE TYPE	NAME	ABBREVIATION	SYMBOL	COMMENTS
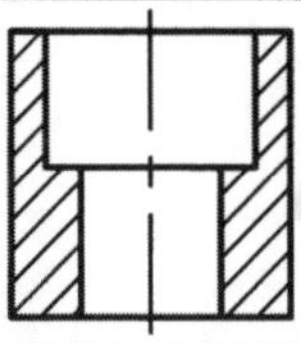	COUNTERBORE	CBORE OR "CB"	⌴	BIG FLAT BOTTOM HOLE OVER SMALLER HOLE. DIA. AND DEPTH MUST BE GIVEN.
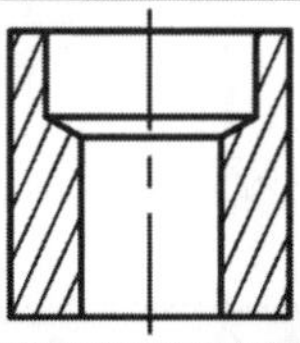	COUNTERDRILL	CDRILL OR "CD"	⌵	BIG DRILL CHASES SMALLER DRILL. DIA. AND DEPTH TO SHOULDER MUST BE GIVEN.
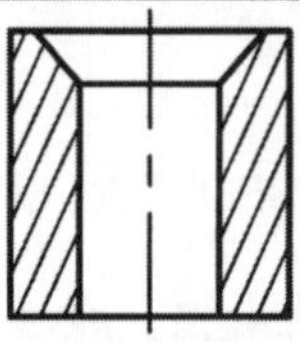	COUNTERSINK	CSINK OR "CSK"	∨	82° POINT ANGLE IS COMMON SO NORMALLY ONLY SCREW HEAD DIA IS REQUIRED.
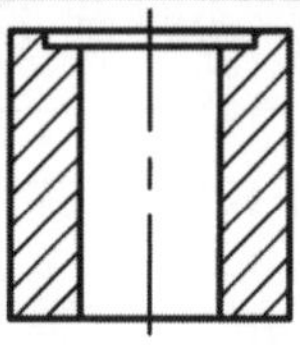	SPOTFACE	SFACE OR "SF"	⌴	BEST TO GIVE DESIRED FINISHED DIAMETER AND OMIT ANY DEPTH REFERENCE.
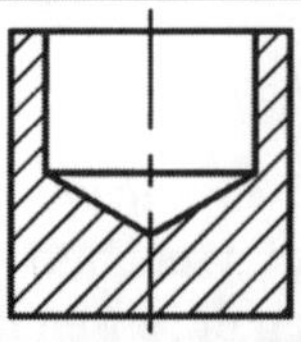	BLIND HOLE	NONE	⌵	HOLE DIAMETER AND DEPTH TO HOLE SHOULDER IS GIVEN.
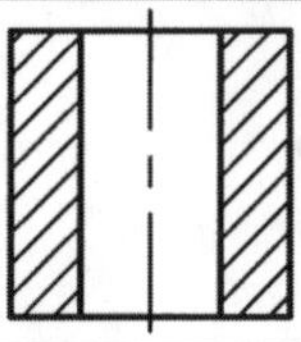	THROUGH HOLE	THRU	NONE	NORMALLY ASSUMED TO GO COMPLETELY THRU UNLESS DEPTH IS GIVEN.

Figure 4.22 Examples of common hole types and symbols.

TYPICAL SPOTFACE NOTES

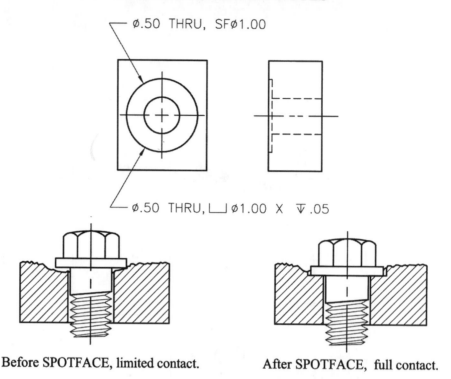

Before SPOTFACE, limited contact. After SPOTFACE, full contact.

Figure 4.23 Spotfacing does not usually require a depth specification as it should only clean an uneven surface deep enough to allow full contact with a bolt or washer.

TYPICAL COUNTERBORE SPECIFICATIONS

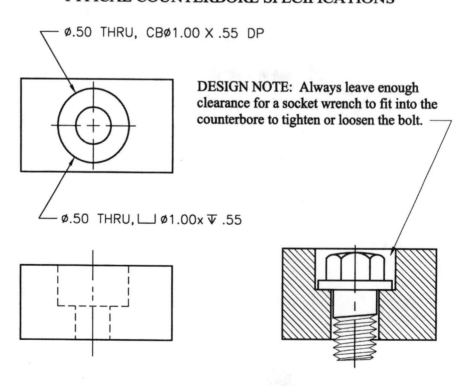

DESIGN NOTE: Always leave enough clearance for a socket wrench to fit into the counterbore to tighten or loosen the bolt.

Figure 4.24 Counterbores are used to keep bolt or screw heads below the finished surface. They require both a diameter and a depth dimension which are based on the size of the bolt head. Examples of both symbols and text notations are shown.

TYPICAL COUNTERDRILL SPECIFICATIONS

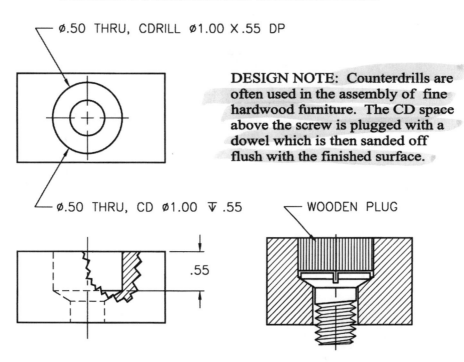

Figure 4.25 Counterdrills are used to keep bolt or screw heads below the finished surface on both wood and metal. They require both a diameter and a depth dimension (measured to the shoulder) which are based on the size and type of the bolt head. Examples of both symbols and text notations are shown.

TYPICAL COUNTERSINK SPECIFICATIONS

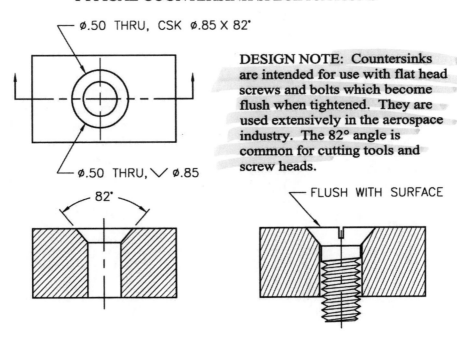

Figure 4.26 Countersinks are used to keep bolt or screw heads flush with the finished surface on wood, metal, and plastic parts. They require a diameter to match the screw head size and the 82° shank angle (shown above for reference only) is very common among cutting tools and fasteners.

TYPICAL BLIND HOLE SPECIFICATIONS

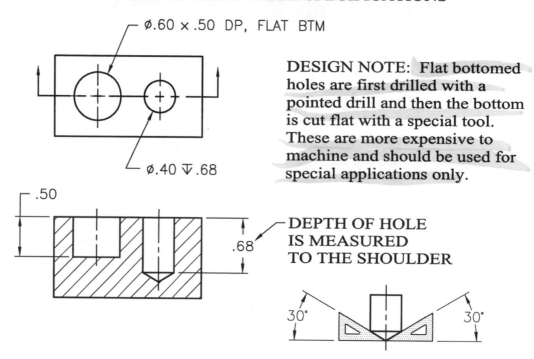

Figure 4.27 Blind holes may have flat or conical bottoms. The conical angle is not specified since it results from the angle of the drill point. It is important to know that this angle (ideally 120°) may vary from time to time as the drill gets dull and has to be sharpened.

TAPPED HOLE SPECIFICATIONS

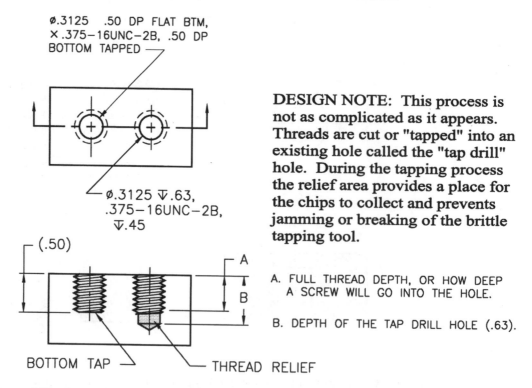

Figure 4.28 Tapped holes are specified with the tap drill size and depth. The thread note is then added. Threads may be cut all the way to the bottom (difficult process) or may have a relief area below them, which is preferred for speed and economy.

Specifications and Dimensioning   81

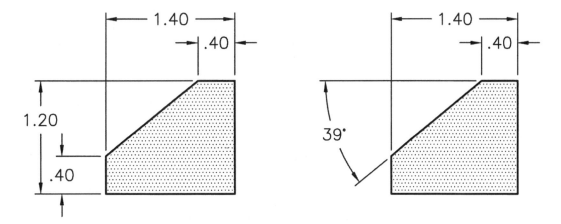

Figure 4.29 Dimension angles with coordinates or with one vertex location and the angle in degrees.

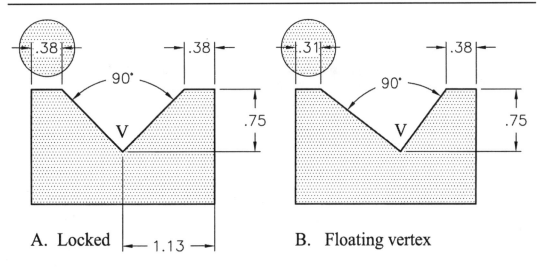

Figure 4.30 Clarify your design intent by locking down floating points like this vertex (V) with both "X" and "Y" dimensions.

Figure 4.31 Chamfers break away sharp edges and may be added for safety. They also help guide the part into a mating assembly. They may be specified with one distance and an angle or with two distances.

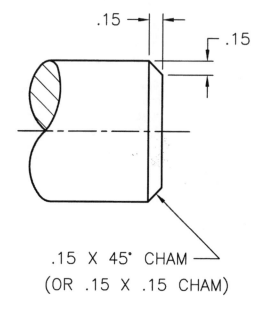

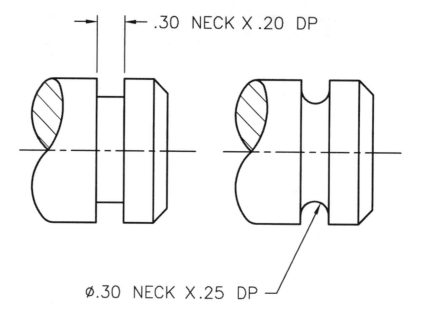

Figure 4.32 Necks are sometimes called grooves. They may have a flat bottom or a rounded bottom and require both depth and width dimensions. They are commonly used for "O" ring seals and for parts that are turned on lathes.

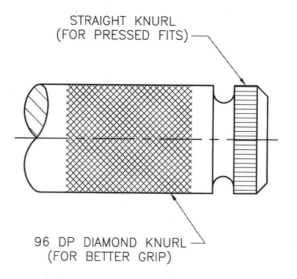

Figure 4.33 Knurling (pron. nerling) is a process of rolling patterns onto cylindrical parts to make them easier to grip. The lines or diamonds are pressed into the base part without removing any material. A common size diamond pattern (96) would place 96 small diamonds around the circumference of a 1.0" shaft.

Techniques for dimensioning in very small places are shown in **Figure 4.34**. Dimensioning of slots may be done with several methods as shown in **Figure 4.35**.

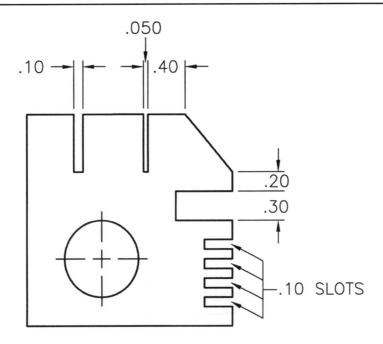

Figure 4.34 Several methods of dimensioning in very small places are shown above. Sometimes very small slits have very lengthy specifications.

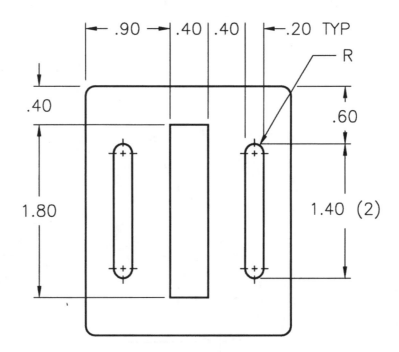

Figure 4.35 Several methods of dimensioning slots are shown above. When the ends are tangent, only the letter "R" is given since it is equal to half of the width (.20/2 = .10R).

Repetitive features can be described with a combination of notes and dimensions as shown in **Figure 4.36**. Locating holes may be done with coordinates and dimensions between centers. This method and the technique for using folded dimension lines for large radii are shown in **Figure 4.37**.

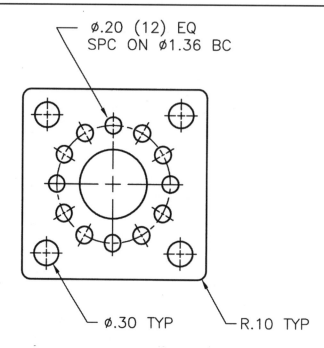

Figure 4.36 For dimensioning repetitive features a combination of dimensions and notes is used. A number in parenthesis (12) indicates how many are required. The word TYPICAL (or TYP) indicates that all shown are the same.

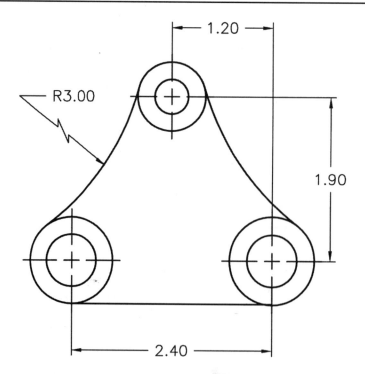

Figure 4.37 Holes should be located to their centers. Multiple holes should be located from center to center. Large radii can be folded to fit on the page.

When dimensions on the drawing are all stated in millimeters, you should place on "SI" symbol near the title blocks. The specifications for making the "SI" symbol are shown in **Figure 4.38**.

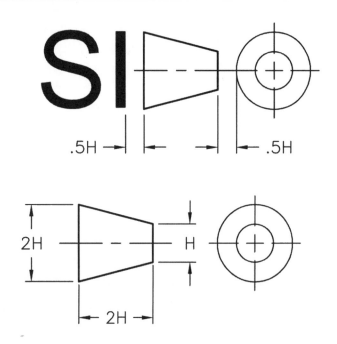

Figure 4.38 Dimensions for construction of the third-angle projection symbol used with metric drawings. "H" indicates the text height. The letters "SI" must be made bold. (ASME Y14.3M)

CHAPTER 4—Practice Quiz: Dimensioning

True or false: Place "T" or "F" in the space to indicate if the statement is true or false.

_____ 1. Placing dimensions at a "T" joint insures the clearest and most accurate size description.

_____ 2. The rules for dimensioning are based on the ASCE nationwide standards.

_____ 3. The recommended text height for dimension text and numerals is .125" or 3mm.

_____ 4. The two methods for positioning dimension text on a three view drawing are trilateral and aligned.

_____ 5. Arrowheads can be drawn either open or closed on pencil drawings.

_____ 6. The size of an arrowhead should have a length that is three times its width.

_____ 7. The first row of dimensions should be spaced a minimum of four text heights (about .50") away from the edge of the part.

_____ 8. Multiple features on the same drawing, like fillets and rounds, should be dimensioned with a general note.

_____ 9. Inside rounded corners between surfaces are known as "rounds."

_____ 10. The distance between the second and third row of dimensions should be two text heights or about 6mm.

_____ 11. The extension line should be offset from the feature it is dimensioning by one half of a text height (about $1/16$").

_____ 12. A drilled hole should be dimensioned in its circular view with its radius.

_____ 13. A leading zero must be included for metric dimensions that are less than 1mm long.

_____ 14. Extension lines are permitted to cross each other, but are not allowed to cross dimension lines.

_____ 15. Dimension and extension lines must be easy to identify so they are made black and thick like visible lines.

_____ 16. Cylinders should be dimensioned with their diameter in their circular view.

_____ 17. The symbol for a diameter is represented by a circle with a "D" in its center.

_____ 18. A counterbored hole has tapered sides to fit with flat head screws.

_____ 19. The depth of a blind hole is measured to the deepest point of the drill recess.

_____ 20. Unlike chain dimensions, baseline and ordinate dimensions minimize tolerance buildup.

Review of Dimensioning Rules

Test yourself by analyzing the two view drawing below and matching the mistakes shown with the dimensioning rule number that is violated.

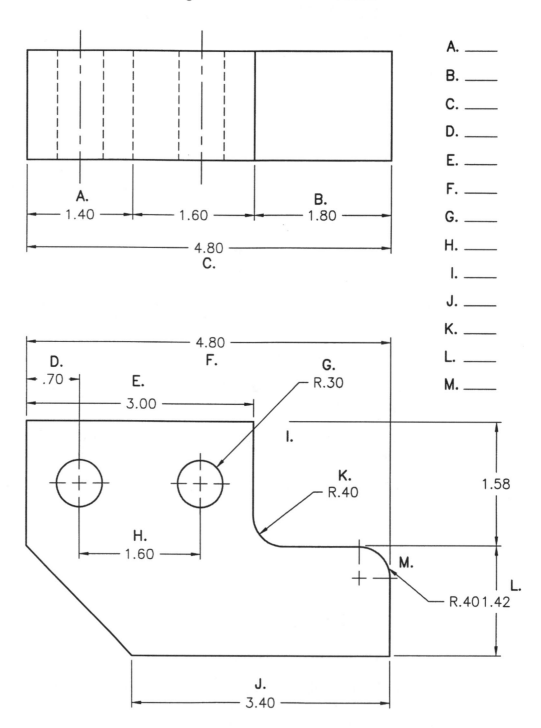

A. ____
B. ____
C. ____
D. ____
E. ____
F. ____
G. ____
H. ____
I. ____
J. ____
K. ____
L. ____
M. ____

FIG. 1

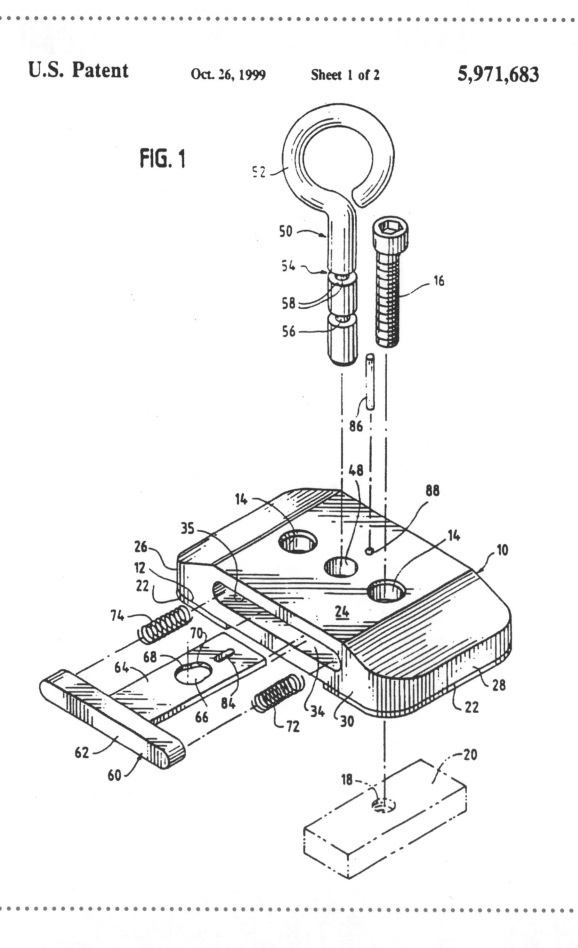

Screw Threads and Fasteners

Machines or other mechanisms made of metal are assembled or held together by welding, rivets, screws, bolts and nuts. Since welding and rivets are intended to be permanent, the use of bolts and screws is very popular because they allow for dismantling and re-assembly for service or adjustment. The earliest known forms of screws were used in ancient Egypt to lift water from the Nile for irrigation. This application was similar to the sketch shown in **Figure 5.1** where a large wooden auger fitted inside a tube was dipped into the river and revolved to form an early pump. Not bad for 3000 years ago when all they had to work with was a little wood and some bronze. In 1415, Leonardo da Vinci designed and built tools used to cut threads, both internal (taps) and external (dies) to be used on some of his ingenious designs. In 1798, the United States inventor, Eli Whitney, best known for inventing the cotton gin, became known as the "father" of the principle of interchangeable parts. His concept of using interchangeable parts was first recognized when he set up a factory to manufacture 10,000 muskets within 28 months for a huge government contract. **(Figure 5.2)** The entire concept of interchangeable parts is based on using precisely made components held in place with bolts, nuts and screws that allows for dismantling and re-assembling. At this time most bolts had to be fitted to one nut, and they were sold only as a set. In recent years, however, the size and form of screws and bolts have become standardized for both the U.S. and the metric systems and they are freely inter-changeable within their classification.

Just for the record, we should state the difference between a bolt and a screw. Most people believe that if you use a screwdriver to tighten the head, then it must be a screw, and that bolts are always bigger than screws. This is not necessarily so, as a screw can have a hex head and a bolt could have a

Figure 5.1

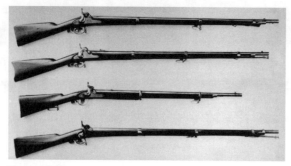

Figure 5.2

slot for a screwdriver. The defining difference is simply that if it has a nut to attach to its end, it is called a bolt (as in bolts and nuts), but if it is screwed into a tapped hole on a machine or assembly and does not require a nut, then it is technically a screw. Engineering society has gotten far away from these original definitions and it's probably not worth the effort to try and retrain them. We will just go with the flow for our use of the terms and assume that big, hex head threaded fasteners are bolts, and smaller ones applied with screwdrivers are screws.

Basic Definition

Mechanical engineers classify the screw as a simple machine since it moves an item forward and backward along an axis by sliding it along a thread. The most basic definition of a screw would be, an **inclined plane** wrapped around a cylinder or cone. The cylinder would be called a **body** and the raised profile on the outside would be called the **threads**. **(Figure 5.3)** The centerline of a cylinder is called the **thread axis**.

As the screw is revolved around the axis, the load (or nut) moves forward or backward along the threads. The distance from the top of one thread to the top of the next thread is called the **pitch**. The distance that the bolt advances with one full revolution is called the **lead** (pronounced "leed"). In most cases, the pitch is equal to the lead. This means that if the threads are $1/10$" apart,

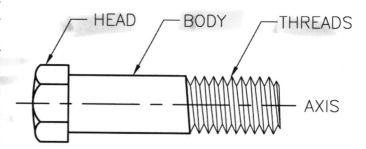

Figure 5.3 A common bolt and its parts. The body is sometimes called the shank.

one full revolution will advance the load $1/10$". On occasion, it is necessary to have a double or triple pitched thread. This allows the load to move twice or three times as far with one revolution. The size and strength rating of a screw, the type of head it has, and the lead specified are all factors to be considered by the designer when selecting the best fastener for a specific application.

The study of bolts, nuts and fasteners has become very scientific and high tech with today's precision measuring and manufacturing machines. The World Wide Web **(WWW)** is filled with reference sites of technical data about screws and fasteners. You need only to search for the key words: bolts, screws, threads, "thread tables" or go directly to the manufacturers like "Americanscrew.com" or "Boltscience.com." You will find data on types, sizes, applications, and tests for yield strength, all very helpful for practicing engineers. No need to reinvent the wheel, it's all been calculated and standardized for us.

Thread Terminology

Almost everywhere we look in our daily lives, we see assemblies held together with bolts and screws. Because of their importance to the designer, it is essential to know the terminology associated with bolts and screws. **Figure 5.4** shows the definition of the nomenclature of the typical screw thread. **Figure 5.5** shows the typical thread notes for a U.S. screw. Among the **most important terms** to understand are those of the **major diameter** and the **pitch**. Given these two bits of information, an engineer will be able to match bolts and nuts and specify the threads needed for a design. You can go into any hardware store and ask for a box of "quarter-twenty" bolts and matching nuts, and the clerk will probably ask for the basic information, "what head type, and how long do you want them." All you need to know is that you need a quarter inch diameter with 20 threads per inch, which is a pitch of $1/20"$. With the U.S. system the **complete note** also indicates if it is coarse or fine, tight or loose, internal or external, or right hand or left hand threads.

Thread Notations in U.S.A.

It is always assumed that the threads are right handed (nut spins on clockwise as you look down the threaded end) unless noted "**LH**" at the end of the thread note for left hand threads (Ex: .75-10UNC-2A-LH). These are desired for applications where the spinning or friction of the application would tend to loosen the bolt. A good example of this application is for bicycle pedals, where the right crank has left hand threads and the left crank has right hand threads. While riding, the pumping of the pedals could unscrew one of them with continuous counter-clockwise friction and vibration. Take a look sometimes and you should see an "R" stamped on the flat end of the pedal screw that you push with your left foot and an "L" on the one you push with your right foot. This is a good example of matching the design intent with the user environment by providing screws that will be continually tightening themselves instead of working loose.

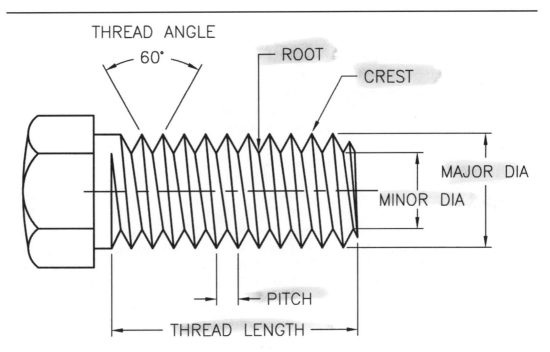

Figure 5.4 The nomenclature of threads. Sometimes a pitch diameter is required. It is the average between the major and minor diameters.

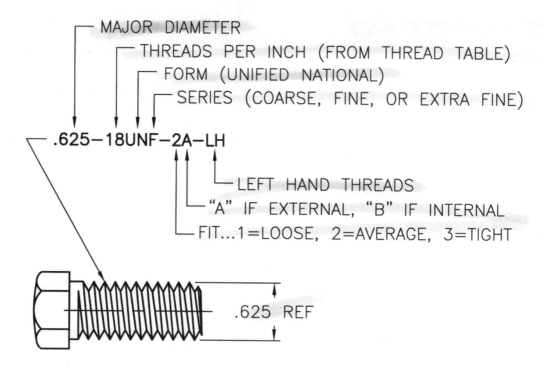

Figure 5.5 Typical thread note used in the U.S.A. They are always considered right hand threads unless the designation "LH" is attached to the end of the note.

Metric Thread Notation

In comparison with the U.S.A. system, the metric system (SI) notation used by most of Europe and the Orient, is much shorter and simply states the **diameter** and **pitch**. If you returned to the hardware store and asked for a box of "**M6 X 1**" bolts (pronounced "m 6 by one") the clerk would again ask how long and what head type? "M" tells the clerk that it is metric, "6" indicates that the diameter of the bolt is 6 mm and the "1" is the pitch. You should recall that pitch is the distance from the crest of one thread to the crest of the next thread. Although there are many other variables for designating metric threads like length of engagement, strength designation, left hand, etc., the metric system is still much simpler to use than the U.S. system. Most applications only call for a simple thread note like "M12 X 1.75." **(Figure 5.6)** Learn the terms and know where to find the thread tables for size references on the WWW or in reference books like *Machinery's Handbook* or this text. **(Appendix A)** In the global engineering environment today, the successful designer will know how to work effectively with both systems.

Thread Symbols

When making design sketches of threaded fasteners, engineers worldwide use three forms of symbols to represent the threads: **Simplified**, **Schematic**, and **Detailed** as illustrated in **Figure 5.7**. Of these, the one that is traditionally used most often is the schematic thread, followed by the simplified. In **Figure 5.7**, it is easy to see that the simplified thread could be confused for a piece of hollow pipe. If not properly drawn, the schematic could be mistaken for hatch symbols or knurling on the outside of the part.

Screw Threads and Fasteners 95

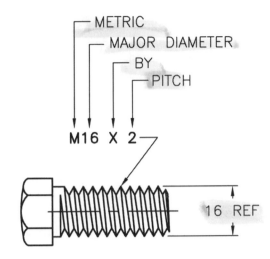

Figure 5.6 Typical metric thread note. Class of fit designation and "LH" may be attached to the end of the note if required for special applications.

The detailed thread is clearly the best communicator, but it also requires the most labor when drawn by hand. Note that the depth of threads, or root line for the simplified and schematic is even with the chamfer at the end of the bolt. For schematic pitch distance use approximately twice this depth. Remember, these are symbols only, not scale drawings, so don't get bogged down trying to count the number of threads needed for the drawing. Imagine how difficult it would be to represent a thread like ".25-28UNF-2A" as it would appear to scale with 28 root lines and 28 crest lines (56 total) crowded into one inch of threads. Definitely a place for symbols . . . not scales.

In today's design arena, with CAD systems being used almost exclusively to produce finished drawings, there are very few reasons not to use detailed symbols for threads. Most CAD systems allow for the use of multiple copies and rectangular arrays that can place detailed threads on a bolt as quickly as schematic threads. Also, most of these systems have parts libraries that include standard parts like bolts, nuts, and screws that are already drawn and simply have to be plugged into the drawing. If these are not available, then you should create template files, so that you would only have to draw the bolts once and use copies in the future.

When it comes to drawing threads, always remember that clear communication is the key factor for sketching, and realistic threads may be unnecessary as the thread note has greater importance. The sketch itself is not necessarily to scale or correctly

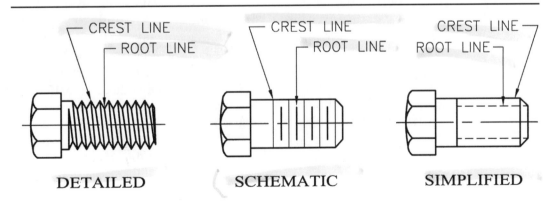

Figure 5.7 Thread symbols are sketched as simplified, schematic, or detailed. For most CAD detailed looks best. Simplified is best suited for very small diameters. Schematic uses bold, visible lines for roots and thin lines for crests.

proportioned, but that is unimportant. The sketch simply communicates that a bolt, nut, or screw is required for that assembly and the thread note details what kind and how big. In other words, it is best to sketch threads with schematic symbols, add an accurate thread note, and finish them with detailed symbols using CAD.

Internal Threads

Holes that are threaded are called tapped holes. The holes are first drilled to the **tap drill size** that comes from charts like Table 2 in **Appendix A** or ones that accompany the machinist's tap and die sets. The diameter of the tap drill is roughly 75% of the major diameter of the thread. If the tapped holes do not go through the part, both a thread depth and a tap drill depth must be provided in addition to the thread note. We indicate internal threads with hidden lines on drawings that represent the **crest** and **root** of the thread as shown in **Figure 5.8**. It is preferred to have the leader point to the visible circle in its round view. If it is not possible to point to the circular view, the leader should point to the crest line in its hidden view as seen in **Figure 5.8**. Because of the short lines used for detailed threads, it is sometimes difficult to show them as hidden, so the simplified works well for this application.

Square and Acme Threads

Square and Acme thread forms are used for power transmission and applications like jacks, clamping devices like C-clamps, and swivel chairs where greater strength is required. They are similar in appearance but the acme has tapered sides on its threads and requires a few more lines to draw as detailed. These would definitely be some to draw as schematic or simplified unless using a CAD system where they are fairly simple to create. The procedure for drawing square threads is shown in **Figure 5.9** and for drawing acme in **Figure 5.10**. These types of bolts and nuts are not as readily available from vendors and must be custom made or special ordered. Naturally, they cost quite a bit more than the "sharp V" or unified forms, so this needs to be factored in while creating the design.

Types of Bolt Heads

There are many types of bolt and screw heads available for use by the designers. The type selected depends on many factors such as surface contact, flush fits, power equipment used for installation, and the clearance around the head from adjacent parts. Some of the more popular types of bolt and screw heads are shown in **Figure 5.11**. Most often, these bolt heads are only sketched and then specified with notes. It is important to understand the principle of the construction of the bolt head as shown in **Figure 5.12** for a standard square head and a hex head bolt. It is also important to understand the principles of these heads in order to make good sketches. If you have to draw them often, drafting templates are available that make them quick and easy to draw. All of these bolt and screw head types are available in both inch and metric forms and there are virtually hundreds of possible combinations. It is a good idea to check your supply catalogs first, find the size and style that fit your design idea, and then adapt your design to use the vendor bolts whenever possible. Bolts and screws can be bought in bulk for a few cents each, but if they have to be customized or made to special order for an odd size that was specified, they could cost several dollars apiece and eat away potential profits.

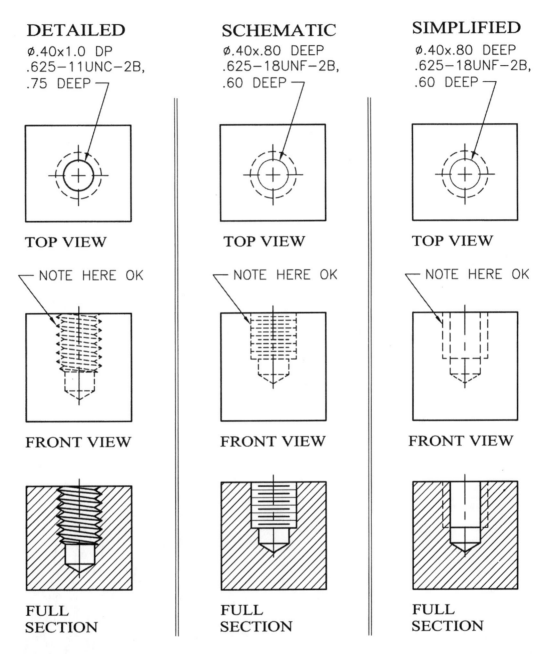

Figure 5.8 Internal threads require the tap drill size and depth as well as the thread note. The key to determining tap drill sizes is "**75**" since the **tap drill diameter is 75%** of the thread diameter and the **thread depth is 75%** as deep as the tap drill. This is measured from the last complete thread to the shoulder of the tap drill hole.

DRAWING SQUARE THREADS

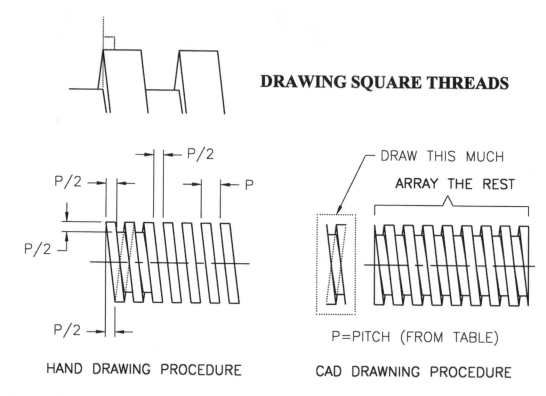

Figure 5.9 Square threads are drawn using the pitch as a key. The **width** of the thread top, the **depth** of its root, and the space to the next thread, are all **one half** of the **pitch**. The angle line from the top thread and root junction are established as shown with the dotted lines.

DRAWING ACME THREADS

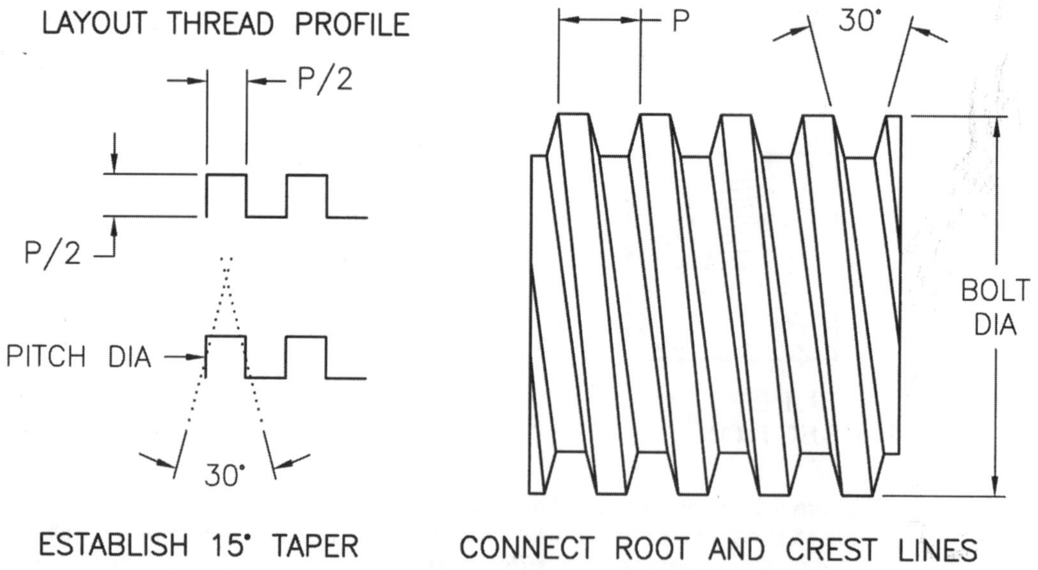

Figure 5.10 Acme threads are more common than square threads with a variety of threaded shafts and nuts available from industrial suppliers.

Screw Threads and Fasteners 99

COMMON TYPES OF BOLT AND SCREW HEADS

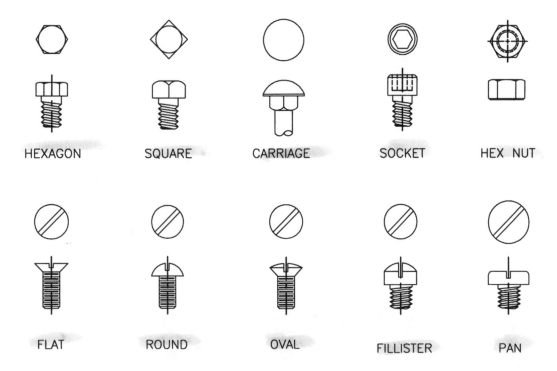

Figure 5.11 The designer needs to know the names and advantages of using each type of bolt or screw. A few of the more common ones are shown here. In just one industrial catalog (mcmastercarr.com), there are more than 4200 combinations of bolt heads, thread types, and sizes listed in stock. By looking under their proper names they are filtered down to the specific type you need for your design.

CONSTRUCTION OF HEXAGON HEADS

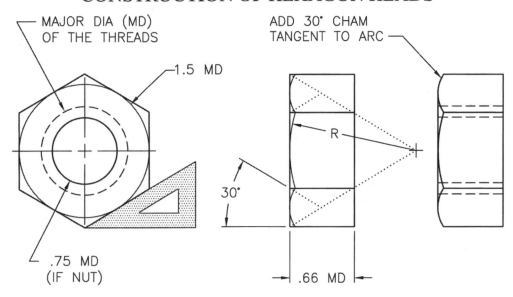

HAND DRAWING PROCEDURE

Figure 5.12 Hex head nuts and bolts are used repeatedly on assemblies. The size ratios for drawing them are shown here. Note that key dimensions are referenced to the major diameter of the thread (MD).

CHAPTER 5—Practice Quiz: Threads and Fasteners

_____ 1. Hardware suppliers classify bolts and screws into two broad classes which are _____.
A) English (U.S.) and metric   B) Fine and extra fine   C) Chrome and dull
D) Tight or loose fitting   E) None of these

_____ 2. If a thread note ends with the letters "3B," it is _____.
A) Metric   B) A loose fit   C) Internal   D) Left hand   E) None of these

_____ 3. The thread note, ".375-16UNC-2B" would have a pitch of _____.
A) .375"   B) .16"   C) .20"   D) $^1/_{16}$"   E) Can't tell without a thread table

_____ 4. The letters "UNC" in a thread note mean _____.
A) Union Certified   B) Use Normal Center   C) Units Numerically Controlled
D) Unified National Coarse   E) None of these

_____ 5. The thread note, ".375-16UNC-2B" would be what kind of fit?
A) Loose   B) Average   C) Tight   D) Interference   E) Can't tell without thread table

_____ 6. The root lines on a schematic thread symbol should always be drawn _____.
A) Thin, like centerlines   B) Wide, like visible lines   C) Alternate, thick and thin
D) Vertical   E) None of these

_____ 7. A thread labeled "M12x1.5" would be _____.
A) 12" long   B) 12mm long   C) 12mm in dia.   D) 1.5 mm in dia.   E) None of these

_____ 8. The distance a bolt advances into a threaded hole with one full revolution is called _____.
A) Spinout   B) Runout   C) Advance   D) Spin Depth   E) None of these

_____ 9. The part of a bolt between the head and the threads that is not threaded is called the _____.
A) Shank   B) Waist   C) Neck   D) Shin   E) None of these

_____ 10. A bolt requiring 100 threads could be drawn quickest using the CAD concept of _____.
A) Copy   B) Multiple copy   C) Array   D) Advanced copy   E) None of these

(11–15) Fill in the blank. Identify each head type shown below with its correct name.

11. _____   12. _____   13. _____

14. _____   15. _____

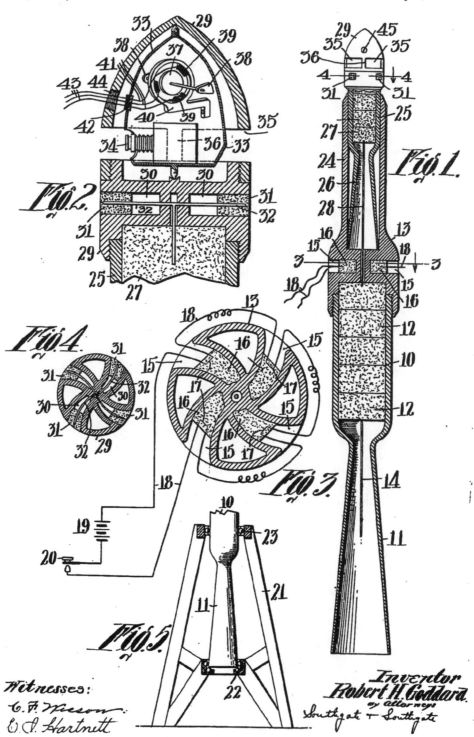

Special Types of Views

In addition to the standard orthographic views used by engineering designers, there is sometimes a need for additional views to clarify the design. Two types of views, the **section view** and the **auxiliary view**, are frequently used. Section views remove a portion of the part's outside surface to reveal some interior detail that would be confusing or otherwise invisible if shown with hidden lines. Section views don't actually remove any physical material but are more like x-rays. The area included in the "x-ray" is shown with a cutting plane. Auxiliary views are used to show the true shape of an inclined surface that would otherwise appear distorted or foreshortened in the orthographic views as seen on the base of the crank mounting post. **(Figure 6.1)**

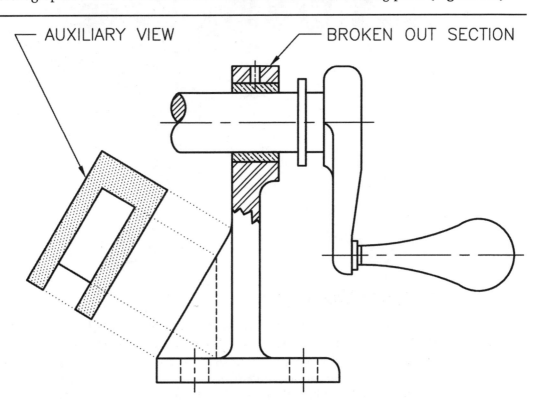

Figure 6.1 The crank assembly shown above uses both a section view and an auxiliary view to illustrate additional details.

Chapter 6

Sectioned Views

Most of the time, all you have to do to create sectioned views is to convert the hidden lines to visible and add crosshatching. Like pictorials, hidden lines are normally omitted when they lie behind the crosshatching. Sectioned views are classified into 6 types:

1. **Broken Out**
2. **Full**
3. **Half**
4. **Revolved**
5. **Removed**
6. **Offset**

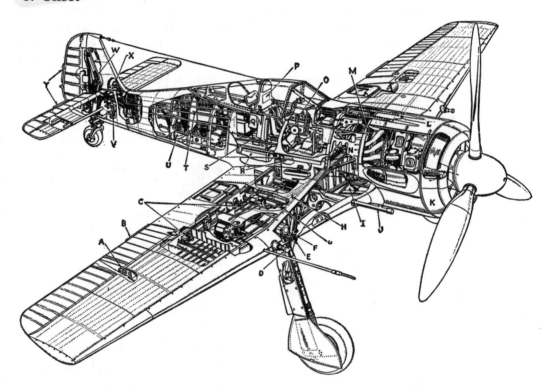

1. Broken Out Section

Comparisons between regular orthographic and the broken out, full and half sections are shown in **Figure 6.2**. Most likely the fastest of all sections to draw, the **broken out section** simply removes a random portion of the outside of the part with an irregular shaped cutting line to reveal some interior detail. The feature being revealed determines the depth of the cut. A good example of this would be like biting into an apple. This removes the outside part of the skin revealing the inside pulp and perhaps some of the core. The broken out is probably the simplest and one of the most frequently used type of sectioned view. **(Figure 6.3)** The lines separating the sectioned portion of the view from the unsectioned portion are called break lines. Different shapes require different treatments called conventional breaks as seen in **Figure 6.4**.

Special Types of Views   105

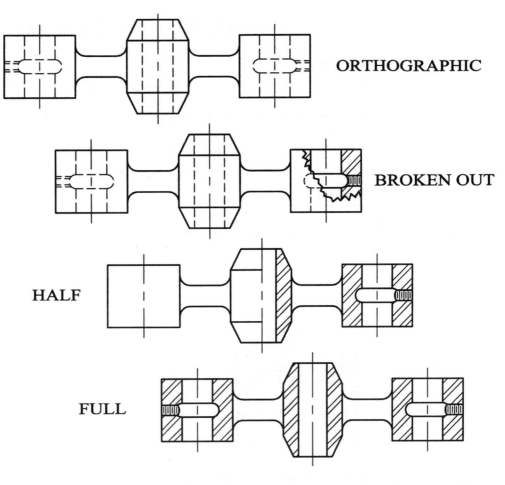

Figure 6.2 Section views remove an imaginary portion of the exterior surface of a view in order to show some interior detail that is unclear or complicated with hidden lines. They are not needed for simple views.

Figure 6.3 A broken out section removes a random "chunk" of the outside surface to any desired depth to reveal the interior details. These are probably the most easily understood sections for persons who have no training in engineering design.

CONVENTIONAL BREAKS

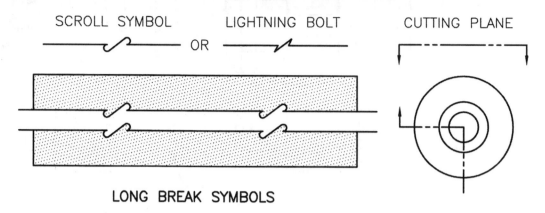

LONG BREAK SYMBOLS

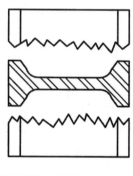

SHORT BREAK LINES

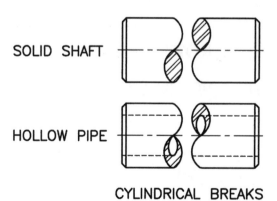

CYLINDRICAL BREAKS

Figure 6.4 Separation lines showing where an object has been broken apart are called "conventional breaks." This practice allows extremely large parts to be shown on the page at a reasonable size.

2. Full Section

The **full section** passes an imaginary knife fully across the object to reveal the interior detail as shown in **Figure 6.5**. The different materials revealed inside are shaded with hatch lines, sometimes called crosshatching. These hatch patterns actually represent different types of materials (like iron or brass) that the parts are made from.

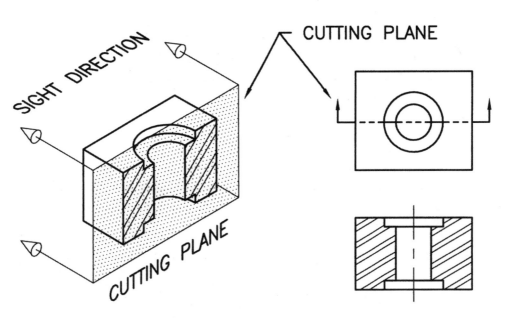

Figure 6.5 A full section cuts completely across the part and removes one half. The arrows on the imaginary cutting plane point to the side that is being viewed.

Examples of 6 of the frequently used patterns are shown in **Figure 6.6** and a complete list as defined by the ANSI, is shown in **Appendix B**. Of these patterns, the one for cast iron is an all purpose or generic pattern used for materials not found in the ANSI files. The international ISO material patterns are not compatible or even similar, but still define how the parts interface with each other. In this respect, sectioned views in both systems are frequently used to define the shapes of internal parts and how they function with each other. **Figure 6.7** shows how a nutcracker would appear as a fully sectioned assembly view and permits a functional analysis to easily be performed. This analysis can show if the ram will engage the anvil with a range of motion as intended by the designer, and if the handle provides enough leverage. When sectioned parts are adjacent to each other, the crosshatching should be altered by the angle or scale to make each part distinctive and easier to visualize.

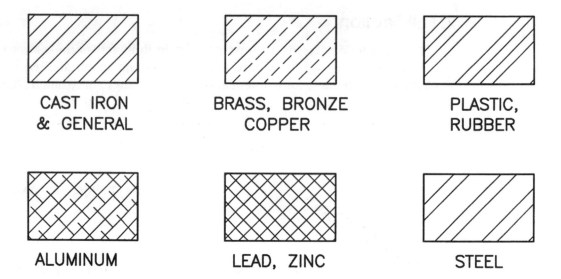

Figure 6.6 Six of the common hatch patterns are shown at full scale above. Hatch lines should be thin like centerlines and spaced approximately 1/16" to 1/8" apart. For additional patterns see Appendix B.

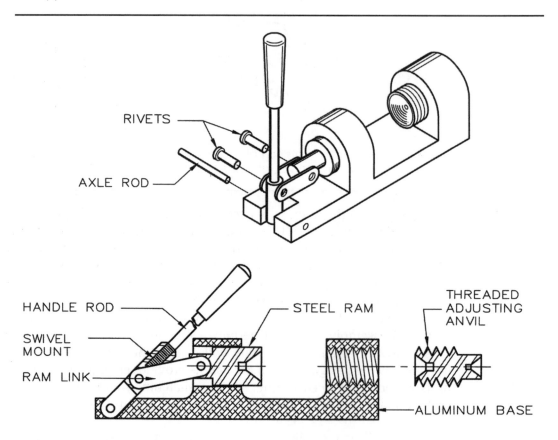

Figure 6.7 Fully sectioned assembly view of a nutcracker. The ram link is shown in front of the assembly for functional analysis.

3. Half Section

The **half section** is used when the part is symmetrical and can result in a considerable savings of time by only drawing hatching on one half of the part. The cutting plane bends 90° and removes $1/4$ of the object, thus revealing half of the object as an outside view and half as an inside view. Aircraft designers frequently use this type of section to show both inside and outside details of an aircraft. **(Figure 6.8)** The break lines separating the hatched area from the unhatched area are shown as either VISIBLE or CENTERLINES as illustrated in **Figure 6.9**. Centerlines are preferred for cylindrical shaped parts and visible for rectangular objects.

HALF SECTION

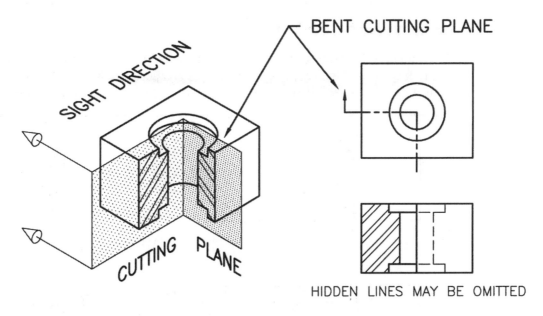

Figure 6.8 A half section cuts halfway across the part and only removes one fourth. These are very useful on symmetrical objects like aircraft, boats, and automobiles. Hidden lines may be omitted on the unhatched side to give an unobstructed outside view.

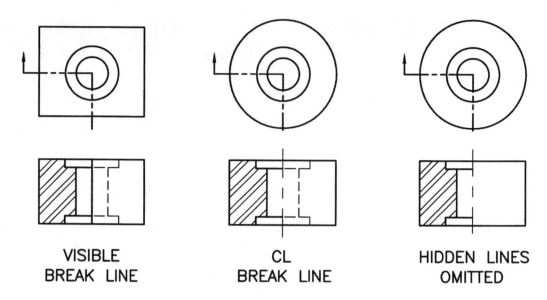

Figure 6.9 Show the break line between the sectioned and unsectioned portions with a visible or centerline. The centerline is preferred for round parts.

4. Revolved Section

Revolved sections are used when space is limited, an object is extremely long, or it makes a transition in cross section such as a baseball bat. A section at the end of the bat where it would strike a ball is larger than the gripping end. **(Figure 6.10)**

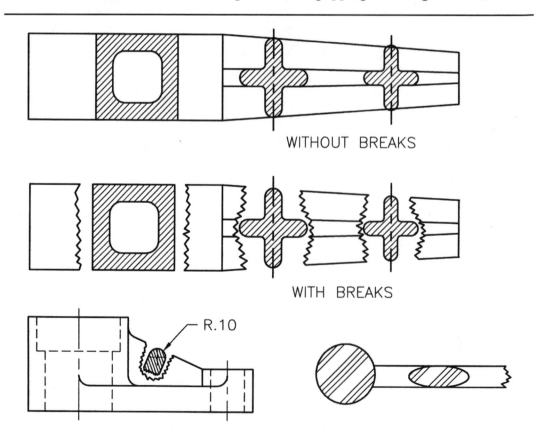

Figure 6.10 Revolved sections are crosshatched on the object. They may be shown with or without the break lines, but break lines can provide extra space for dimensioning when needed.

Special Types of Views   111

5. Removed Section

A **removed section** is also used for long or transition parts but is best used when there is not enough space for a revolved section. This also allows room for adding dimensions to the cross section. The resulting section view is connected to the parent view with a cutting line. **(Figure 6.11)** For very large assemblies, the removed view might even be placed on another page. Views placed on another page should be designated with a reference note such as "Section AA, BB, CC, etc." or some other alpha character as shown in the figure.

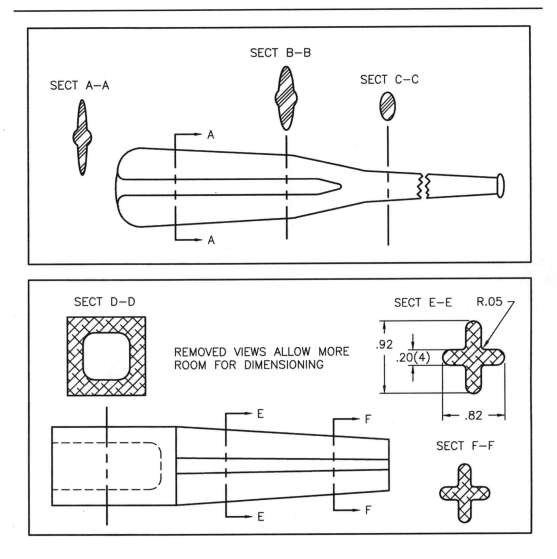

Figure 6.11 Removed sections are very useful when space is too limited on the part for a revolved section. They are connected to the parent view with a cutting plane line. They can also be labeled and placed at other locations on the page or within the set of drawings.

6. Offset Section

The **offset section** is a full section except the cutting plane makes 90° bends across the part instead of a straight line. These 90° bends are made across the part in order to pass through critical features that would otherwise be missed by the cutting plane and remain unseen. **(Figure 6.12)** Since cutting planes are not really removing any material, no lines are generated by these bends in adjacent views.

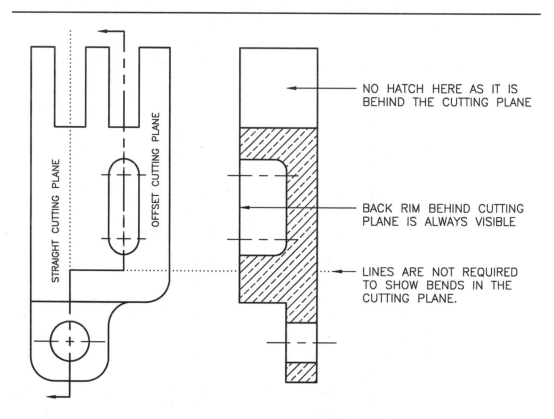

Figure 6.12 Straight cutting planes can bypass important details. Offset sections make 90° bends to pass through them as necessary to correctly describe the features. Bends in the cutting plane do not create lines in the sectioned view.

General Rules for Sectioned Lines

When drawing hatch lines, it is best to use a 4H pencil and make very thin black lines such as centerlines. When placing the crosshatching on the part, it is important that they are not mistaken for screw threads. In this regard, they should always be made at a contrasting angle, never perpendicular to an outside edge. The most popular angles used for hatch lines are 30°, 60° and 45°. The spacing between normal full sized hatch lines is approximately $1/16"-1/8"$, but vary with the size of the part. Naturally, larger parts can have the hatch symbols spaced much wider but smaller parts may have to omit them totally because of limited space. For extremely thin parts, it is recommended that the material be shaded solid and described with a label instead of trying to squeeze in hatch symbols. When two sectioned parts are touching each other, the angle or scale of each hatch pattern should be different to provide contract and separation. **(Figure 6.13)**

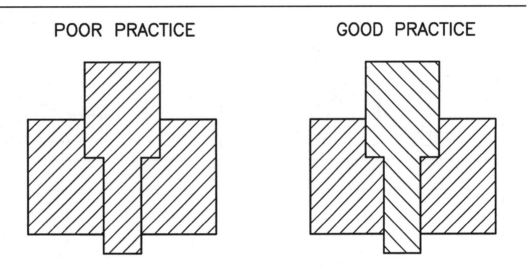

Figure 6.13 Alternate the hatch angle or scale (or both) to provide contrast and distinction between close parts.

Aligned Parts

If there are an odd number of radial features that do not lie on the cutting plane, they must be theoretically aligned in the section view. **(Figure 6.14)** This is a conventional practice rather than true orthographic projection. It allows the features to remain true distance from the outside edge and convey the design intent more accurately.

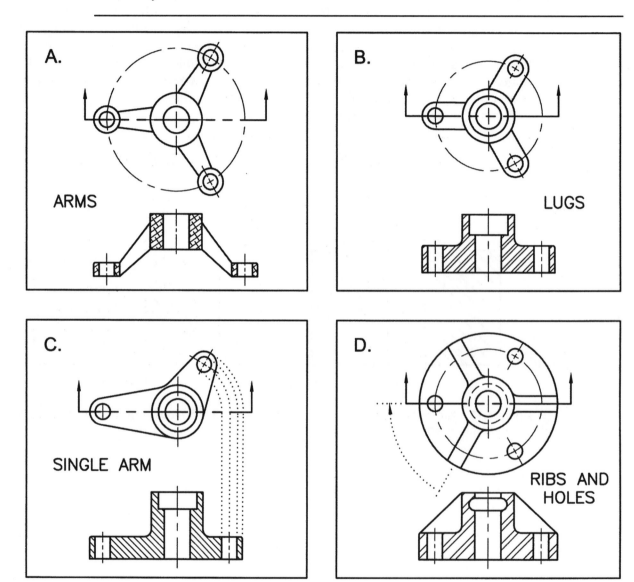

Figure 6.14 Conventional practice for aligning lugs, ribs, arms, and holes.

Parts Not Sectioned

To insure clarity and save time, some features of parts do not get crosshatched even though the cutting plane passes through them. These include RIBS, WEBS, SPOKES and most vendor items. Vendor items include BOLTS, NUTS, WASHERS, SCREWS AND SHAFTS. **(Figure 6.15)**

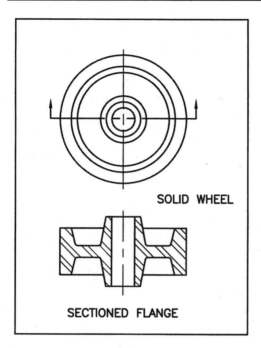
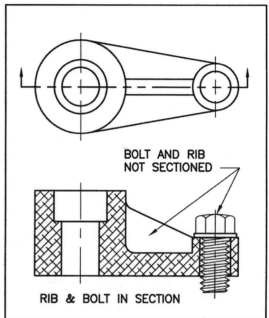
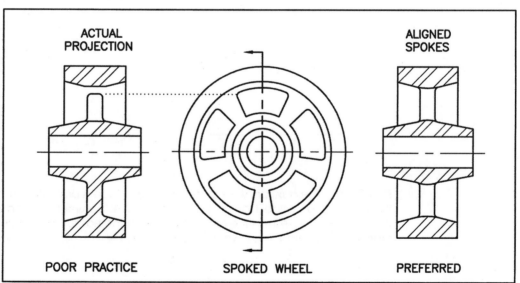

Figure 6.15 **Spokes, ribs, webs,** and most **vendor parts** are not crosshatched in the sectioned view. Common vendor parts include bolts, nuts, pins, washers, and screws.

Auxiliary Views

A simple definition of auxiliary is "additional support," as in an auxiliary engine on a sailing yacht. In engineering graphics, auxiliary views give additional support to the 6 principle views in order to show the true size and shape of inclined surfaces. **Figure 6.16** illustrates an object that is not true size in the top view. To calculate the area of the inclined surface, the true size distances must be known. In orthographic projection, an imaginary glass plane was placed parallel to the front/back, left/right or the top/bottom to establish normal views. For the *auxiliary* view, the glass plane is placed parallel to edge view of the inclined surface as in **Figure 6.17** and is called a reference plane. It is critical that the designer transfer the proper dimensions into the auxiliary view. The simple rule is, if the auxiliary view is projected from a view such as a front view with the dimensions of height and width given, then the designer must go to the adjacent top or a profile view to get the missing dimensions of depth, as shown on the workshop roof in **Figure 6.18**.

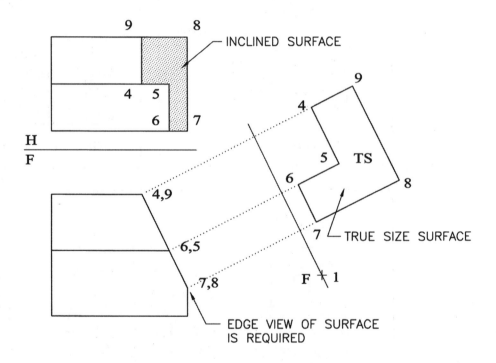

Figure 6.16 The auxiliary view of the oblique surface is projected from the view where it appears as an edge. Note the difference in size between the shaded view and the true size. "H/F" represents the fold line between the horizontal and frontal planes. "F/1" represents the fold line between the frontal and first auxiliary planes.

Special Types of Views 117

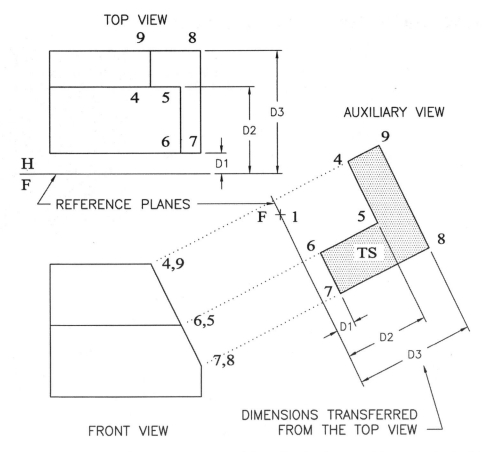

Figure 6.17 Since the auxiliary view is projected from the front view which provides height and width, the depths (D) must be transferred from the top view.

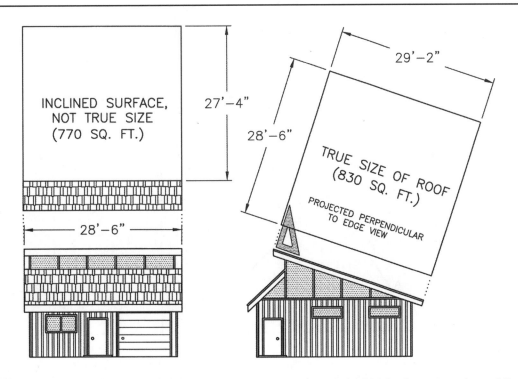

Figure 6.18 The workshop auxiliary view is projected normal (at 90°) to the edge view of the roof plane. This gives a true size view of the roof as compared to the top view which is foreshortened.

Descriptive Geometry Concepts

Auxiliary views can become quite complicated with all of the projections and measurements involved. The study of auxiliary views evolved from an area of engineering design called descriptive geometry. Descriptive geometry allows the designer to establish true sizes and shapes, angles of inclination, compass bearings, and point views of lines and planes in space. **Figure 6.19** shows the methodology required to get the true length of a line. Any line parallel to the projection plane is shown in its true length in the view opposite the reference plane. **Figure 6.20** carries this one step further to establish the point view of a line. Once the point view of a line is found, it automatically makes it easy to find the edge view of a

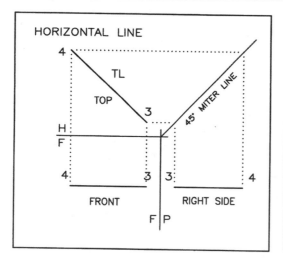
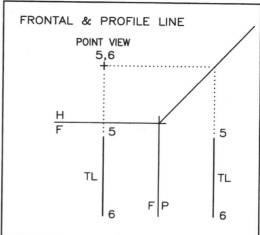
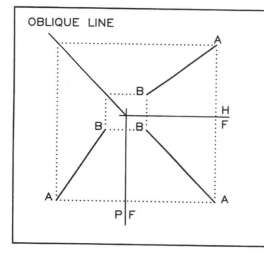
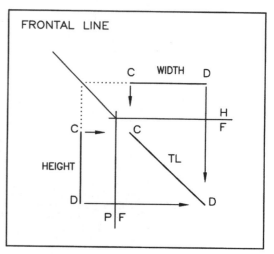

Figure 6.19 In descriptive geometry, lines can be projected from any two views to obtain a third view. If the line is parallel to the projection plane, it will project as a true length (TL).

Special Types of Views 119

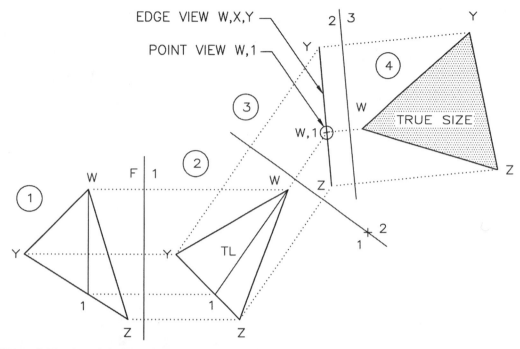

Figure 6.20 A point view can be projected from a true length line (2), which establishes an edge view (3), which finally projects to a full sized view (4). Calculations for the perimeter and area can now be made.

plane. Once the edge view of a plane is found, the next projection will be the true size of a plane revealing needed information such as the area, volume, or mass. True size surfaces can be established and connected to form the flat pattern of an object. These are used extensively for designs in aerospace, packaging, and HVAC industries. **(Figure 6.21)** Descriptive geometry concepts are still used extensively by civil engineering for infrastructure of cities such as sewer, water and electric utilities, road, bridge and dam construction. These concepts are the historical method of solving problems. With some additional training, most of these concepts can now be performed on CAD systems. Even when using the computer, knowing the principles of these projections will aid greatly in knowing how to solve problems with CAD.

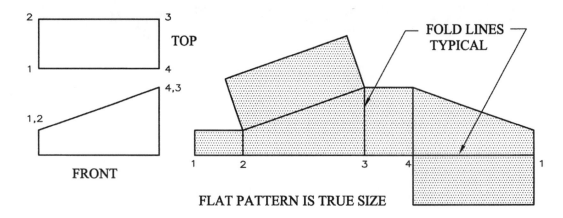

Figure 6.21 True size surfaces can be established and connected to form the flat pattern of an object. These are used extensively for designs in aerospace, packaging, and HVAC industries.

CHAPTER 6—Practice Quiz: Section and Auxiliary Views

____ 1. Which of these is not a standard type of section view?
A) Full   B) Associated   C) Half   D) Removed   E) None of these

____ 2. The type of section view whose cutting plane makes 90° bends across the part in order to show additional details is called a ____ section.
A) bent   B) zigzag   C) crisscross   D) offset   E) none of these

____ 3. The cutting plane line is a line that has ____ .
A) long segment, two dashes, and thick weight   B) long segment, one dash, and thick weight   C) long segment, two dashes, and thin weight   D) depends on the hatch pattern selected   E) none of these

____ 4. The recommended spacing between crosshatched lines like cast iron should be spaced approximately ____ apart.
A) $1/4"$ B) $1/16"-1/8"$ C) $1/8"-1/4"$ D) $1/4"-1/2"$ E) none of these

____ 5. The pencil lead best suited for drawing crosshatch symbols is ____ .
A) 4H   B) 4B   C) 2B   D) 8H   E) none of these

____ 6. Parts that are not sectioned, even though the cutting plane passes right through them, include all of the following except ____ .
A) fillets   B) ribs   C) bolts   D) spokes   E) none of these

____ 7. The type of section view that shows both the exterior and interior details of a part at the same time is ____ .
A) broken out   B) half   C) removed   D) rotated   E) all of these

____ 8. An auxiliary view is always projected from ____ .
A) the top view   B) the front view   C) an edge view   D) a true size view   E) none of these

Refer to the hatch patterns shown below to answer questions 9–12.

____ 9. The crosshatch pattern for Steel is ____ .
A)   B)   C)   D)   E) none of these

____ 10. The crosshatch pattern for Aluminum is ____ .
A)   B)   C)   D)   E) none of these

____ 11. The crosshatch pattern for Plastic is ____ .
A)   B)   C)   D)   E) none of these

____ 12. The crosshatch pattern for Brass is ____ .
A)   B)   C)   D)   E) none of these

Fig. 1

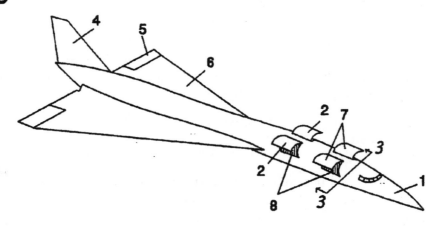

Fig. 2

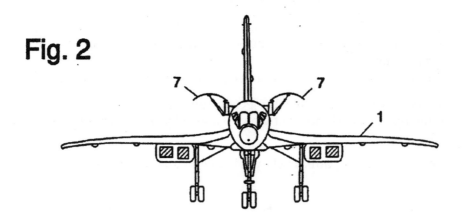

Fig. 3

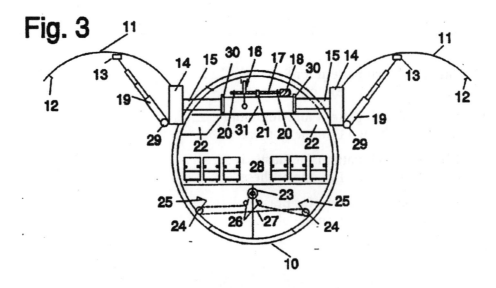

Tolerances in Design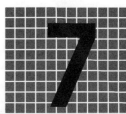

The concept of using tolerances in design can be very confusing or very easy . . . let's keep it very easy. The first example: Let's say you have a board 3'-0" long and you need it to be sawed off to be only 2'-0" long. You go to a friend who has a saw and ask him to saw off 1'-0" foot of the board. He would take his tape measure and mark off 1'-0" and make the cut. When you measure the board after cutting, it may be slightly shorter than 2'-0" because the friend did not allow for the thickness of the saw blade. Now you return with another board and you communicate your design intent a little more clearly. You say, "I need this board to be exactly 2'-0" long." This time, he will know which side of the line to make the cut. It would be most difficult to make any dimension exact, but this time the chance of having a board exactly 2'-0" long is much better. In the case of lumber, if the saw is outside in the sunlight and weather, the wood is going to expand at the time it is measured. After the cut is made and you return back into the air-conditioned climate, the board will shrink some and it may be .01" or .05" shorter than when it was outside. Just imagine how many boards would be wasted before they actually came up with one exactly 2'-0" long when measured on the inside. It is because of examples like this that the concept of tolerance and precision evolved.

Tolerance is defined as the permissible deviation from the basic size of the part. In our example, imagine that you returned with the board and said this board needs to be 2'-0" long but can be $1/16$" longer than 2'-0" or $1/16$" shorter than 2'-0". This would give your friend some leeway in setting up his saw and would produce an accurate part much faster than trying to make one exactly 2'-0" long. Let's take the concept one step further, what if you returned with the board and say this board can be 2'-0" long and it can be $1/4$" shorter or longer. In other words, it would be ±.25" (plus or minus .25") in length. It would be much easier for the person to cut this board and much faster than the previous requirements. The lesson to be learned is the **larger the tolerance**, the **easier** it is to manufacture and generally, the **cheaper** it is to make the part.

Expressing Part Sizes—Part sizes can be expressed in three ways:

1. *Nominal Size* is the descriptive sizes of a part like a 2" x 4" board which is actually only 1.5" x 3.5".
2. *Basic Size* is the ideal size to which tolerances are applied. It is sometimes called design size.
3. *Actual Size* is the measured size of an actual manufactured part. These should be measured at a room temperature of 68° while the part is at rest.

EXPRESSION OF TOLERANCES

Tolerances are expressed in three forms:

1. **Bilateral form**—Finished part can be larger or smaller than the basic size.
2. **Unilateral form**—Dimensions can only go in one direction.
3. **Limit form**—Any value between the upper and lower limit is acceptable.

As the name implies, **bilateral** means the finished part can be larger or smaller than the basic size. If a part is 1.5'-0" long, a bilateral dimension could be added saying 1.50" +/−.01. This means the part could be .01" larger or .01" smaller than 1.50". **(Figure 7.1)**

Most companies express **general tolerances** for their drawings as bilateral tolerances located in the title block. Dimensions that are shown without tolerances on the drawing apply these general tolerances. An example of a typical general tolerance note is:

UNLESS SPECIFIED TOLERANCES ARE:

.X   ± .05
.XX   ± .025
.XXX ± .005

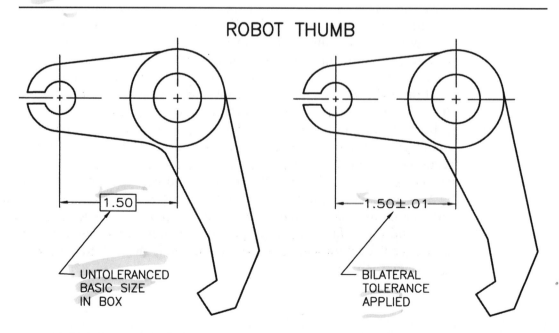

Figure 7.1 Basic sizes before tolerancing are placed inside of a rectangular box. A bilateral tolerance is applied to this basic size.

Tolerances in Design

Unilateral dimensions are used when the dimension can only go in one direction. As compared to the bilateral, the dimension would be stated as 1.50" +.00/-.01 (plus zero, minus .01) meaning the part cannot be any larger than 1.50" but it can be smaller by .01". "Uni" (meaning 1) says the part can expand or contract in one direction only. This is particularly useful in mating parts. If there is a 1.0" shaft that is supposed to go into a 1.0" hole, the hole could not be smaller than 1.0" but it could be larger. On the other hand, if the shaft is expressed as a unilateral dimension, then it could be smaller than 1.0" but not larger than 1.0". **(Figure 7.2)**

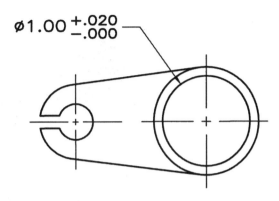

Figure 7.2 A **unilateral** tolerance is applied to this hole so that it will never be less than 1.0" but may be .020" larger to insure fit with shaft.

The **limit form** dimension (Ø1.495"-1.505") is probably the easiest for the machinist to understand and is the most flexible for them to use. In this system, the upper limit of the dimension is 1.505" and the lower limit is 1.495" so any value that falls between this upper and lower limit is acceptable. **(Figure 7.3)** When using limit form, the difference between the upper limit and lower limit on a single part is called the part's **tolerance**. When written in a single line, write the lower limit first. When the figures are stacked, the upper (larger) value is always on the top. The limit form lends itself very well to high production runs in the form of "Go-Nogo" gauges like the one shown in **Figure 7.4**.

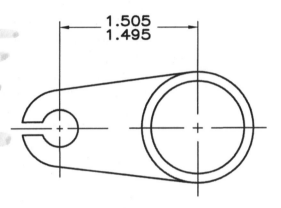

Figure 7.3 A **limit form** tolerance is applied which defines the range of permissible size between the holes.

These inspection devices have each end ground to size to match the upper and lower limits as specified. If the "Nogo" part fits into the hole, then it is a bad part and must be reworked or discarded. Having these gauges available at the machine allow the machine operator to track the wear on the cutting tool and have it sharpened when it first starts to produce faulty parts.

Let's not forget the basic concept to keep the cost down and the quality up. The easy rule is, "*only use tolerances where tolerances are required.*" If you were building a deck on your patio out of wood, the tolerances would not be nearly as critical as if you were building an airlock on a submarine or a spacecraft. The proper use of tolerances can keep the cost down to assure the parts always fit together, but the improper use can raise the cost of the part to unacceptable levels and may not increase the performance significantly. Albert Einstein said it best when he said, "*The trick is to make the good easy to do and the bad difficult to do.*" Use this concept and there should be very few problems with your future designs.

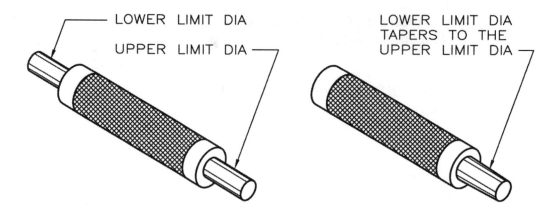

Figure 7.4 A "**go-nogo**" gauge is sometimes used to verify that features such as holes and slots are within limits. Two variations are shown here.

Design for Mating Parts

The main application for tolerancing parts is when parts must mate with other parts. Eli Whitney is credited with being the father of the concept of interchangeable parts. In 1798, he received a contract from the US government to produce 10,000 muskets for the military in only 28 months. He bought a gristmill at Mill Rock, outside of New Haven, Connecticut, and converted it into a musket factory. In the process, he designed his own machines, jigs and fixtures to insure consistency of size and shapes for each part. The machines had to work very accurately as they were operated by relatively unskilled laborers. He invented the milling machine for this operation and except for being computer controlled, they still operate on the same principle today whenever metal is being shaped. Proper use of tolerancing guarantees the parts in an assembly will always fit together. Whitney took several baskets of parts before Congress and demonstrated that the parts could be scrambled and still be assembled as a functional musket. Again, the design intent should be reflected. The design of a garden gate would not require nearly as precise a tolerance as the hatch designed for a deep sea submersible. Two terms associated with the mating of parts are **allowance** and **clearance**.

Clearance is very obvious as it is the loosest combination of the two parts. We see signs on bridges every day saying distances like 16'-6" clearance. If your truck is taller than 16'-6", it will not fit under that bridge. If your truck is only 15'-0" tall, you would have a clearance of 1'-6". Let's take a look at an engineering example using an axle with a Ø1.50" that has to fit into a hole of the same size. If both have a bilateral tolerance of ±.005", the **clearance** (loosest combination of fits), would be determined by subtracting the smallest axle size (1.495") from the largest size hole (1.505"). Thus, the maximum clearance would be .010".

The other variable used when combining mating parts is the tightest combination of fits called the **allowance**. The allowance is determined by subtracting the largest shaft from the smallest hole. In this example, the largest shaft would be 1.505" and the smallest hole would be 1.495". This would result in a force fit since the shaft is larger than the hole giving us a negative allowance of –.010". **(Figure 7.5)**

Tolerances in Design   127

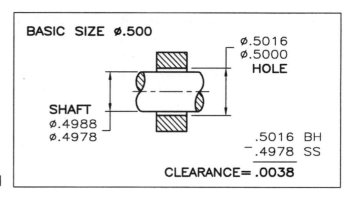

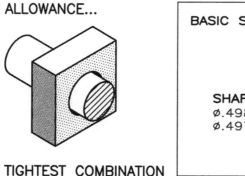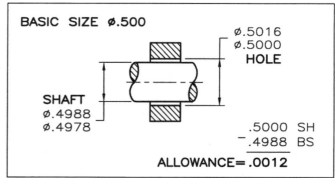

Figure 7.5 Clearance is the loosest combination between two parts... BIG HOLE (BH)-SMALL SHAFT (SS) and **allowance** is the tightest combination... SMALL HOLE (SH)-BIG SHAFT (BS). Tolerance values from "Vinson Quick Table" RC6.

Types of Fits

There are various combinations of fitting mating parts together. The most frequently used are:

1. **Clearance fit**—shaft always clears hole
2. **Interference fit**—shaft always hits sides of hole
3. **Transition fit**—shaft may hit or miss the sides of hole
4. **Line fit**—shaft may be same size as hole or smaller

1. The **clearance fit** is a design where the parts will always clear each other. In other words, the hole, under any circumstances, will always be larger than the specified shaft. **(Figure 7.6)**

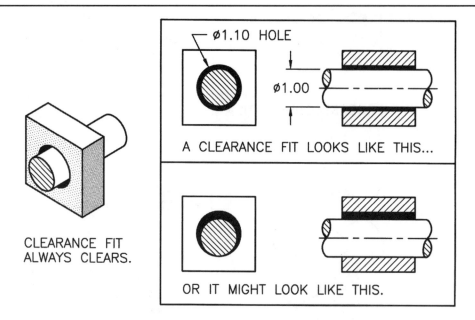

Figure 7.6 A clearance fit between a 1.0" shaft and 1.10" hole. Clearance fits always have a shaft size smaller than the hole size.

2. An **interference fit**, as the name implies, will have a shaft that is slightly larger than the hole. This is sometimes called a force fit because the shaft literally has to be pounded into a hole using force. These are used for high-pressure applications, vacuums, or stud bolts when required. **(Figure 7.7)**

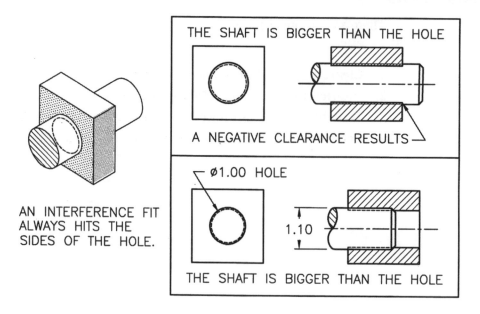

Figure 7.7 An **interference fit** always hits the sides of the hole and must be driven in with force. Because of this it is sometimes called a force fit. For an extreme illustration only, a shaft of 1.10" Dia. is shown forced into a 1.00" Dia. hole.

3. A **transition fit** is a combination of a clearance fit or a interference fit. Some combinations of transition fits will clear while other combinations will interfere. This is because the tolerances specified overlap each other. **(Figure 7.8)** Sometimes, the shaft can be larger than the hole and other times it will be smaller than the hole because of the wide range of tolerance applied to each. You might wonder why such a sloppy design concept would even be used in a precise profession like engineering. The concept of transition fits is used for a type of manufacturing called **selective assembly**. This is where the person doing the assembly of the parts will find that some of the pieces fit too loose or too tight and some are just right. Their job is to find the ones that fit just right and sort the other parts into the tight or loose bins. This works where labor is cheap and results in a considerable savings. The tradeoff, however, is that the parts cannot be easily interchanged in the field since the end user will not necessarily get a tight, loose, or just right fit for their machine. Consequently, this type of fit is popular for production of "throw-away" items with limited life expectancy.

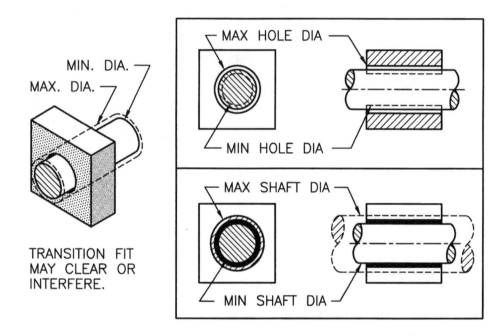

Figure 7.8 A transition fit may clear the hole or hit it due to overlapping tolerances. They are used with a manufacturing process called "**Selective Assembly**" where the best combination of parts are mated together.

4. **Line fits** result when the largest size shaft has exactly the same dimension as the smallest size hole. On a drawing, both hole and shaft edges would be in a straight line. **(Figure 7.9)** These result in zero clearance when this combination is used but the parts will always fit together. When the biggest shaft is aligned with the smallest hole, the two surfaces form a straight line. Zero clearance would not be desirable if the shaft were supposed to rotate, but if the shaft were supposed to be held firmly in place, a line fit would be desirable.

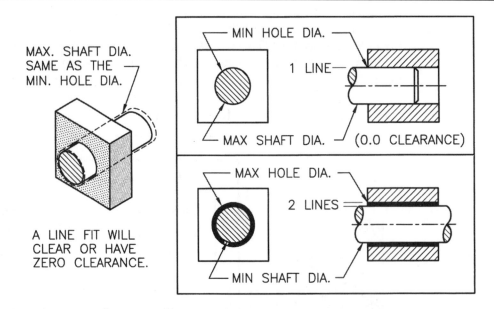

Figure 7.9 The **line fit**, also called a "**clearance locational fit**" may result in either a clearance or a zero clearance, with the shaft and the hole being exactly the same size.

Summary of Tolerancing

The simple rules for tolerancing guarantee that parts will always fit to reflect your design intent. Parts that always clear are called **clearance fits**, parts that never clear are called **interference fits**, and parts that may or may not interfere are called **transition fits**. **Tolerance** is the extreme permissible deviation in basic size of a single part. In order to calculate allowance or clearance, there must be two parts in the design. It is probably safest for the new engineer to use the standardized fit tables as seen in **Appendix A-Table 12**. The values on the tables are based on the nine ranges of fits from tight (RC1) to loose (RC9). Fits for each value are intended for:

1. **RC1—Close Sliding Fits** are used for parts that must assemble without a great deal of play.
2. **RC2—Sliding Fits** are used when the parts need to move and turn easily.
3. **RC3—Precision Running Fits** are used when close fits are needed to move freely but snug.
4. **RC4—Close Running Fits** are used when accuracy and minimum play is needed.
5. **RC5 & RC6—Medium Running Fits** are used for high running speeds.
6. **RC7—Free Running Fits** are used where accuracy is not essential.
7. **RC8 & RC9—Loose Running Fits** are used where wide commercial tolerances are needed.

For convenience, values for some common sizes of tight, medium and loose fits have been calculated in **Figure 7.10**. By using these standard tables, engineers protect themselves somewhat from liability if they use them correctly with sound engineering practice. In lieu of these tables, many companies establish their own systems for fits based on their product needs. For example, an aircraft manufacturer would have very tight specifications for their tolerances as compared to the manufacturer of fence posts.

When the need is justified and tolerances are properly applied, they can save labor as well as money and guarantee precision. If used improperly, they can waste time, money and jeopardize your reputation as an engineer.

Vinson's Quick Reference Table
Common Running & Sliding Fits in Inches

NORMAL SIZE RANGE				Class RC2 Sliding Fit			Class RC4 Close Running Fit			Class RC6 Medium Running Fit			Class RC8 Loose Running Fit	
Fraction	Decimal		Limits of Clearance	HOLE H6	SHAFT g5	Limits of Clearance	HOLE H8	SHAFT f7	Limits of Clearance	HOLE H9	SHAFT e8	Limits of Clearance	HOLE H10	SHAFT c9
1/16	0.0625	MAX	0.10	0.06275	0.06240	0.3	0.06256	0.06247	0.6	0.06350	0.06244	2.5	0.06410	0.06000
		MIN	0.55	0.06250	0.06220	1.3	0.06250	0.06243	2.2	0.06250	0.06090	5.1	0.06250	0.05900
1/18	0.125	MAX	0.15	0.12530	0.12485	0.4	0.12507	0.12496	0.8	0.12620	0.11700	2.8	0.12680	0.12220
		MIN	0.65	0.12500	0.12465	1.6	0.12500	0.12491	2.7	0.12500	0.12350	5.8	0.12500	0.12100
3/16	0.1875	MAX	0.15	0.18780	0.18735	0.4	0.18757	0.18746	0.8	0.18870	0.17950	2.8	0.18930	0.18470
		MIN	0.65	0.18750	0.18715	1.6	0.18750	0.18741	2.7	0.18750	0.18600	5.8	0.18750	0.18350
1/4	0.25	MAX	0.20	0.25040	0.24980	0.5	0.25009	0.24995	1.0	0.25140	0.24900	3.0	0.25220	0.24700
		MIN	0.85	0.25000	0.24955	2.0	0.25000	0.24890	3.3	0.25000	0.24810	6.6	0.25000	0.24560
5/16	0.3125	MAX	0.20	0.31290	0.31230	0.5	0.31259	0.31245	1.0	0.31390	0.31150	3.0	0.31470	0.30950
		MIN	0.85	0.31250	0.31205	2.0	0.31250	0.31140	3.3	0.31250	0.31060	6.6	0.31250	0.30810
3/8	0.375	MAX	0.20	0.37540	0.37480	0.5	0.37509	0.37495	1.0	0.37640	0.37400	3.0	0.37720	0.37200
		MIN	0.85	0.37500	0.37455	2.0	0.37500	0.37390	3.3	0.37500	0.37310	6.6	0.37500	0.37060
7/16	0.4375	MAX	0.25	0.43790	0.43725	0.6	0.43850	0.43744	1.2	0.43910	0.43630	3.5	0.44030	0.43400
		MIN	0.95	0.43750	0.43695	2.3	0.43750	0.43620	3.8	0.43750	0.43530	7.9	0.43750	0.43240
1/2	0.5	MAX	0.25	0.50040	0.49975	0.6	0.50100	0.49994	1.2	0.50160	0.49880	3.5	0.50280	0.49650
		MIN	0.95	0.50000	0.49945	2.3	0.50000	0.49870	3.8	0.50000	0.49780	7.9	0.50000	0.49490
9/16	0.5625	MAX	0.25	0.56290	0.56225	0.6	0.56350	0.56244	1.2	0.56410	0.56130	3.5	0.56530	0.55900
		MIN	0.95	0.56250	0.56195	2.3	0.56250	0.56120	3.8	0.56250	0.56030	7.9	0.56250	0.55740
5/8	0.625	MAX	0.25	0.62540	0.62475	0.6	0.62600	0.62494	1.2	0.62660	0.62380	3.5	0.62780	0.62150
		MIN	0.95	0.62500	0.62445	2.3	0.62500	0.62370	3.8	0.62500	0.62280	7.9	0.62500	0.61990
11/16	0.6875	MAX	0.25	0.68790	0.68725	0.6	0.68850	0.68744	1.2	0.68910	0.68630	3.5	0.69030	0.68400
		MIN	0.95	0.68750	0.68695	2.3	0.68750	0.68620	3.8	0.68750	0.68530	7.9	0.68750	0.68240
3/4	0.75	MAX	0.30	0.75050	0.74970	0.1	0.75120	0.74992	1.6	0.75200	0.74840	4.5	0.75350	0.74550
		MIN	1.20	0.75000	0.74930	2.8	0.75000	0.74840	4.8	0.75000	0.74720	10.0	0.75000	0.74350
13/16	0.8125	MAX	0.30	0.81300	0.81220	0.8	0.81370	0.81242	1.6	0.81450	0.81090	4.5	0.81600	0.80800
		MIN	1.20	0.81250	0.81180	2.8	0.81250	0.81090	4.8	0.81250	0.80970	10.0	0.81250	0.80600
7/8	0.875	MAX	0.30	0.87550	0.87470	0.1	0.87620	0.87492	1.6	0.87700	0.87340	4.5	0.87850	0.87050
		MIN	1.20	0.87500	0.87430	2.8	0.87500	0.87340	4.8	0.87500	0.87220	10.0	0.87500	0.86850
15/16	0.9375	MAX	0.30	0.93800	0.93720	0.8	0.93870	0.93742	1.6	0.93950	0.93590	4.5	0.94100	0.93300
		MIN	1.20	0.93750	0.93680	2.8	0.93750	0.93590	4.8	0.93750	0.93470	10.0	0.93750	0.93100
1.0	1.0	MAX	0.30	1.00050	0.99970	0.1	1.00120	0.99992	1.6	1.00200	0.99840	4.5	1.00350	0.99550
		MIN	1.20	1.00000	0.99930	2.8	1.00000	0.99840	4.8	1.00000	0.99720	10.0	1.00000	0.99350
1 1/16	1.0625	MAX	0.30	1.06300	1.06220	0.8	1.06370	1.06242	1.6	1.06450	1.06090	4.5	1.06600	1.05800
		MIN	1.20	1.06250	1.06180	2.8	1.06250	1.06090	4.8	1.06250	1.05970	10.0	1.06250	1.05600
1 1/8	1.125	MAX	0.30	1.12550	1.12470	0.1	1.12620	1.12492	1.6	1.12700	1.12340	4.5	1.12850	1.12050
		MIN	1.20	1.12500	1.12430	2.8	1.12500	1.12340	4.8	1.12500	1.12220	10.0	1.12500	1.11850
1 3/16	1.1875	MAX	0.30	1.18810	1.18710	0.8	1.18870	1.18742	1.6	1.18950	1.18590	4.5	1.19100	1.18300
		MIN	1.20	1.18750	1.18670	2.8	1.18750	1.18590	4.8	1.18750	1.18470	10.0	1.18750	1.18100
1 1/4	1.25	MAX	0.40	1.25060	1.24960	1.0	1.25160	1.24900	2.0	1.25250	1.24800	5.0	1.25400	1.24250
		MIN	1.40	1.25000	1.24920	3.6	1.25000	1.24800	6.1	1.25000	1.24640	11.5	1.25000	1.24250
1 5/16	1.3125	MAX	0.40	1.31310	1.31210	1.0	1.31410	1.31150	2.0	1.31500	1.31050	5.0	1.31650	1.30750
		MIN	1.40	1.31250	1.31170	3.6	1.31250	1.31050	6.1	1.31250	1.30890	11.5	1.31250	1.30500
1 3/8	1.375	MAX	0.40	1.37560	1.37460	1.0	1.37660	1.37400	2.0	1.37750	1.37300	5.0	1.37900	1.37000
		MIN	1.40	1.37500	1.37420	3.6	1.37500	1.37370	6.1	1.37500	1.37300	11.5	1.37500	1.36750
1 7/16	1.4375	MAX	0.40	1.43810	1.43710	1.0	1.43910	1.43650	2.0	1.44000	1.43550	5.0	1.44150	1.43250
		MIN	1.40	1.43750	1.43670	3.6	1.43750	1.43550	6.1	1.43750	1.43390	11.5	1.43750	1.43000
1 1/2	1.50	MAX	0.40	1.50060	1.49960	1.0	1.50160	1.49900	2.0	1.50250	1.49800	5.0	1.50400	1.49500
		MIN	1.40	1.50000	1.49920	3.6	1.50000	1.49800	6.1	1.50000	1.49640	11.5	1.50000	1.49250

Calculated from ASME Table of Values.

Figure 7.10 For convenience, values for some common size fits have been calculated and shown in this figure.

Geometric Tolerance

Now that you know about size and position specifications, you need to understand the impact that the part geometry has on mating parts. This is a relatively new field of study since it was only introduced in the 1950s. The formal name for this area of specifications is Geometric Dimensions and Tolerancing (GDT), sometimes called "GeoTol." There are entire books on the topic as well as full semester classes and many companies offer concentrated 40-hour courses. You can find numerous resources and course presenters on the WWW by searching under "GDT" or "GEOTOL." There are even GDT classes online and available on video. This unit will only cover the basics of GDT including the symbols, feature control frames, and general applications.

What is GDT? Specifications written on drawings may state that a part is flat or round, but GDT sets up a format of guides that allow us to specify just how flat, how parallel, how round, etc. These specifications cover every conceivable position a designer can imagine. **Figure 7.11** shows the geometric shapes addressed under each of the following 5 groups.

1. **Form**—straightness, flatness, circularity, cylindricity
2. **Profile**—line profile, surface profile
3. **Orientation**—angularity, perpendicularity, parallelism
4. **Location**—position, concentricity, symmetry
5. **Runout**—circular runout, total runout

It is important to know the shape of each symbol and understand how it is used in feature control frames **(Figure 7.12)** to lock down the part geometry. These notes and

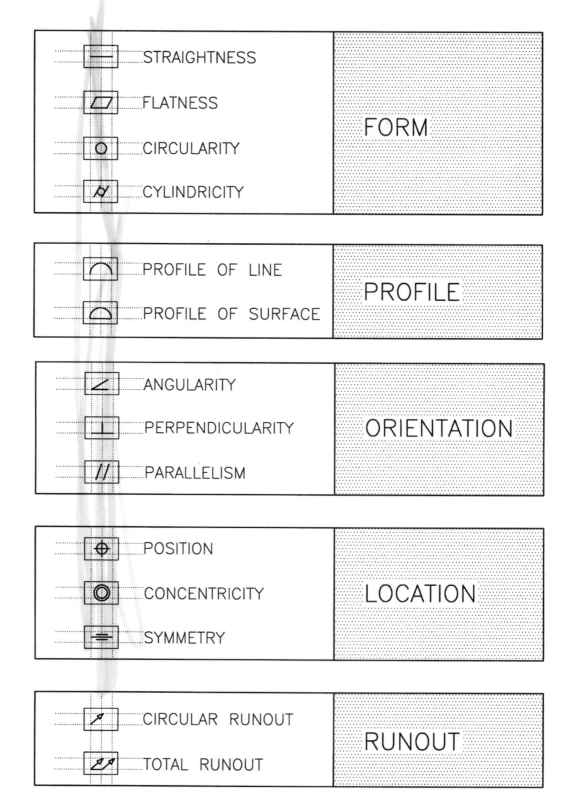

Figure 7.11 Symbols used for geometric dimensioning and tolerancing "GDT." The dotted lines represent text height for estimating size ratios.

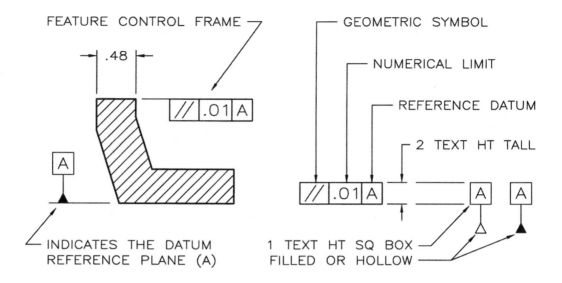

Figure 7.12 The feature control frame ("callout box") is drawn with dimension lines. Its width will vary with the symbols used, but it is always two text heights tall. The datum base is a 60° equilateral triangle, two text heights tall. It may be left hollow, but filled shows up better.

frames can become fairly complicated when you get down to manipulating them to control 6 degrees of freedom. The 6 degrees of freedom refer to 6 possible directions a part can move: up, down, left, right, forward and backward. We will stick to the basics and start with this easy example. Let's say there is a platform like the one shown in **Figure 7.13** and the bottom and top edges should be parallel within .005". This could be done with a leader and a note, but the protocol is to use a feature **control frame** or "**callout box**." Machinists and inspectors are trained to look for these so we might as well learn to use them. The condition of parallel requires at least two surfaces. One surface is the datum (reference plane or baseline) and all other surfaces are referenced to it. The symbol for parallel used in the callout box is the same one that we use in math that looks like a lazy "11" tilted to the right. The callout box shown back in **Figure 7.12** tells us the symbol has to go first, followed by the tolerance value and datum identifier. They are in logical order and they are read just like a sentence "parallel within .01" to datum A." Parallel, perpendicular and angular specifications must always be related to a reference plane (datum) in order to describe their orientation. Some features like flatness or straightness are able to stand-alone since they refer to "form" which is independent of datums. Examples of each of the 14 symbols are shown with their applications in **Figures 7.14–7.16**.

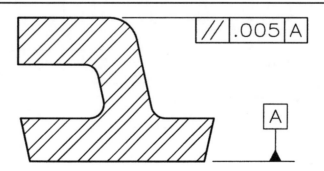

Figure 7.13 The feature control frame can be read like a sentence, "parallel within .005 inches to A."

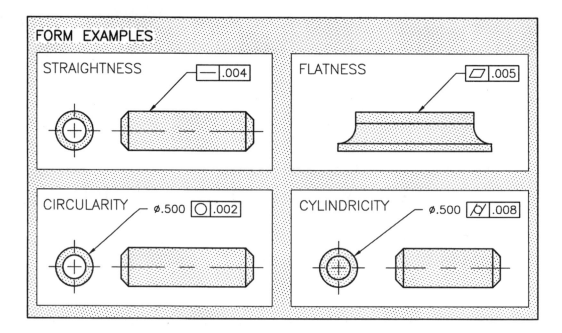

Figure 7.14 Examples of GDT applications for form.

The following figures provide examples of the eleven characteristics of form, orientation, location, and runout for your reference. Modifying symbols shown in **Figure 7.17** indicate if the part is being toleranced from the **Maximum Material Condition "MMC"** or the **Least Material Condition "LMC."** If no modifier is present, it is automatically assumed the values are **Regardless of Feature Size "RFS."** The RFS notation and symbol "S" is no longer used in the US since this assumption is widely accepted and practiced. MMC means that if the dimension limits are shown as 1.500"–1.525", you would base your calculations for GDT on the maximum size of 1.525". LMC would base its values on "1.500"."

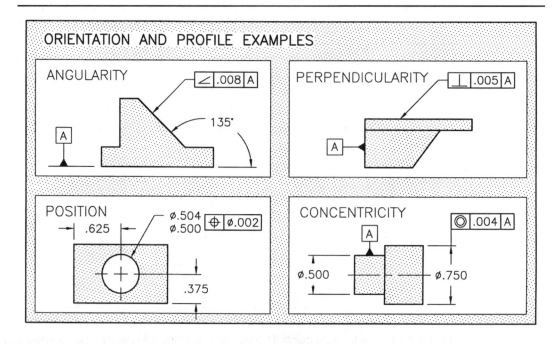

Figure 7.15 Examples of GDT applications of orientation and profile.

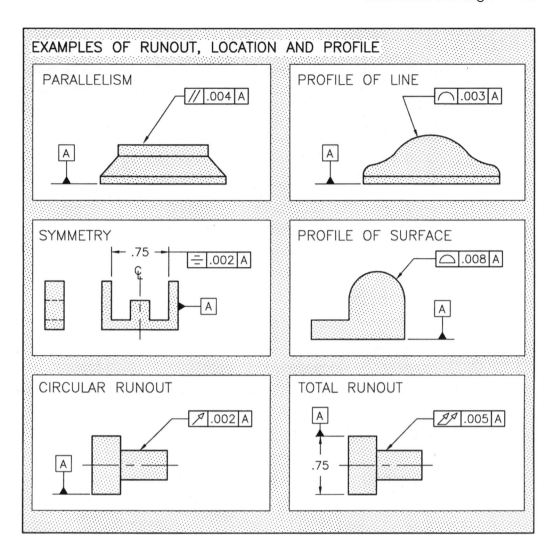

Figure 7.16 Examples of GDT applications for runout, location, and profile.

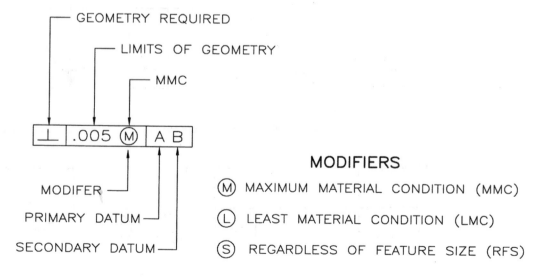

Figure 7.17 Modifiers are often attached to the limiting dimension.

The symbols used in the callout boxes are uniform in the United States but there are still a few minor conflicts with symbols used by the ISO in Europe. The ISO has recently modified several of their symbols to be the same as those specified by the ANSI Y14.5 and a common set of symbols has been proposed. The ISO 9000 initiative will help to standardize these symbols worldwide, as it will be mutually beneficial in this age of global engineering. These symbols are usually drawn freehand but the size ratio to text heights is shown for reference in **Figure 7.18**. Most CAD software has a library of these symbols to use with dimensioning and parametric solid modeling allows you to build geometric constraints into your design.

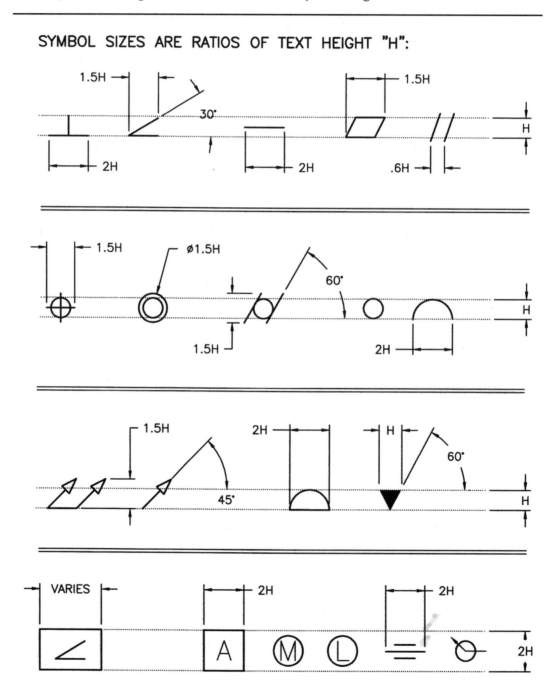

Figure 7.18 Geometric symbols should be drawn to these sizes as per ANSI Y14.5M.

Don't go wild with your geometric tolerancing. Remember to use them only where necessary to control the function of your design. No need to specify a 10-micron finish on a surface that only touches air. Before you start locking down shapes and writing specifications, it is best to acquaint yourself with your company's shop capabilities and limitations. No reason to conjure up designs that can't be economically built.

Also, remember to check the standards of precision used by subcontractors and vendors. It would be useless to cap off a precision assembly with imprecise add-ons that will degrade your engineering design. Finally, when specifying GDT for your design, always consider how they can be inspected and verified. Consistency and repeatability of the inspection processes is critical in producing quality parts.

CHAPTER 7—Practice Quiz: Tolerances in Design

_____ 1. As a general rule, the smaller the tolerance, the cheaper it is to manufacture the part.
A) True   B) False

_____ 2. Allowance may be defined as the _____ .
A) tightest combination between two parts   B) the allowable slack of a single part
C) the loosest combination between two parts   D) tolerance of a single part
E) none of the above

_____ 3. A dimension that is listed as 1.505 ± 005 is a _____ dimension.
A) bipolar   B) binomial   C) bi-directional   D) bilateral

_____ 4. A dimension that has rectangular box drawn around it needs to _____ .
A) be toleranced   B) be squared   C) be placed on a flat plane   D) have the part's corners made square   E) none of the above

_____ 5. Unilateral dimensions and unidirectional dimensions mean the same thing.
A) True   B) False

_____ 6. The value derived from subtracting the lower limit from the upper limit on a single part is _____ .
A) tolerance   B) linear deviation   C) allowance slack   D) axial variation   E) none of the above

_____ 7. In order to save money on part production, it is best to specify individual tolerances on every part.
A) True   B) False

_____ 8. In GDT, the true position of a hole would be best located with a _____ .
A) polar coordinate   B) baseline coordinate   C) chain dimension   D) diameter
E) none of the above

_____ 9. The edge from which GDT dimensions are referenced is called a _____ .
A) baseline   B) datum   C) reference edge   D) binomial edge   E) none of the above

_____ 10. A dimension that appears in parenthesis (.750) is always a _____ dimension.
A) hidden   B) critical   C) reference   D) baseline   E) none of the above

_____ 11. A hole (Ø.500-.526) at maximum material condition would be the _____ .
A) smallest diameter   B) largest diameter   C) average diameter   D) tolerance diameter   E) none of the above

_____ 12. The actual size that you draw GDT symbols is determined by the _____ .
A) size of the text   B) size of the drawing sheet   C) size of the part   D) metric or inch designation   E) none of the above

Chapter 7—Practical Exercises

1. Make a sketch of each GDT symbol shown below and name it on the line below the symbol.

   A. ▱   B. ⌾   C. ⊕   D. ⌀

2. Correct each feature control frame shown below by placing its elements in the correct locations.

   | .005 | ▱ | B | | A | ⌀.003 | ⊕ | | ⊥ | B | .008 |

Refer to the hole and shaft shown below and answer the following questions.

3. What is the hole tolerance? _____

4. What is the shaft tolerance? _____

5. What is the allowance? _____

6. What is the maximum clearance? _____

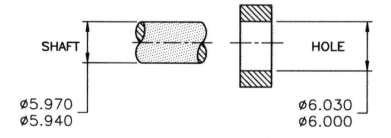

SHAFT
⌀5.970
⌀5.940

HOLE
⌀6.030
⌀6.000

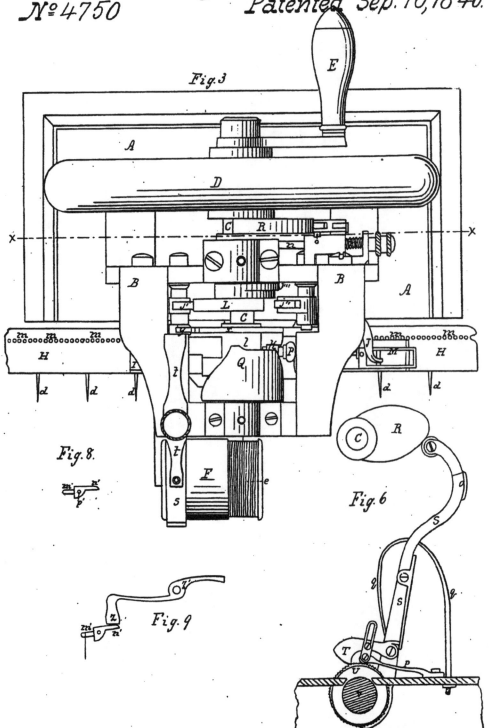

Working Drawings

Until now, we have concentrated on drawing only one part at a time. More complex designs require many parts and views to show how they fit together. All of these parts are combined into one set of **working drawings** that include all of the necessary details to manufacture each part. It must also include an assembly view to show how they all fit together. When the mechanism is made up of many small parts, it is better to put them all on one or more large drawing sheets instead of having each part on an individual sheet. These sheets should include a border, title block and revision zone, and may be customized with logos and industry specific data for the company that is creating the design. The three phases of creating a set of working drawings are:

1. **Layout and Analysis**—Freehand sketches and may include prototyping, patents, destructive testing, etc.
2. **Detail and Specifications for Production**—Includes full details for manufacturing based on data from testing.
3. **Assembly Drawings**—Illustrates how the parts fit together and function as a unit.

Layout and analysis drawings are often based on freehand sketches of conceptual ideas. These may be transformed into CAD drawings or parametric solid models to facilitate prototyping and testing. The results of the tests indicate if the design will be produced or if it needs revising. Patent drawings often are generated in this phase creating a need for documentation and secrecy.

Once the tests are complete and the market analysis is favorable, the drawings will be converted into a detailed set of working drawings. These drawings must contain all of the **details and specifications** necessary for the production and assembly of the product. **(Figure 8.1)** Just imagine how many parts and subassemblies are involved in a set of working drawings for an automobile, aircraft, or even a computer.

An **assembly drawing** is then made, often with copies from the detailed views, to aid the production team in putting the unit together. These views are also used later to show how the unit may be disassembled for service or repair. They are frequently included with the original packing or operating manual provided to the consumer. If a part is missing or broken, the consumer can find the part name and number and order a replacement.

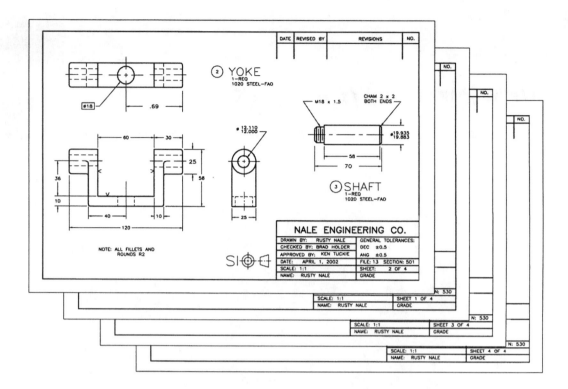

Figure 8.1 A set of working drawings should include all information required to build the parts. Size A3 are sheets shown at reduced scale.

Sheet Layouts

Before the actual drawing begins, the designer makes rough sketches on an engineering pad to estimate the type of views that will be required and the size and placement of each view. From here, a light layout is made on larger sheets to make sure all of the parts will fit in the space. **Figure 8.2** shows a layout that looked pretty good until the dimensions were added and made it look too crowded. Start with large sheets and work at a reasonable scale to show the details clearly, concisely and accurately. Paper costs are by far the cheapest element of the design process so use all you need rather than cramping the views and losing clarity. **Figure 8.3** shows an assembly view where the parts were crowded to fit the space and is not very clear. The "after" treatment shows how much clearer it is when spread out on a larger page. **Figure 8.4** shows the 5 recommended sizes for engineering drawings in both ANSI and ISO sizes. The ANSI system has a base size of 8.5" x 11" (Size A) and when scaled up properly, expands upward to become a large 34" x 44" sheet (Size E). Size "E," when folded correctly ends up as a 8.5" x 11" again which accommodates everything from file folders, mailing envelopes, copy machines, and printers. Large size sheets are very well suited for CAD drawings since the area of interest can be zoomed in for close work and zoomed out to adjust the layout and orientation of parts. Since the cost of large format printers has dropped significantly in recent years, most companies have units that will print up to 36" wide for any length of a roll of paper. The day of the blueprints and diazo prints has passed and the era of the laser printing is upon us. One very nice thing about sending CAD drawings to printers is that the printers can

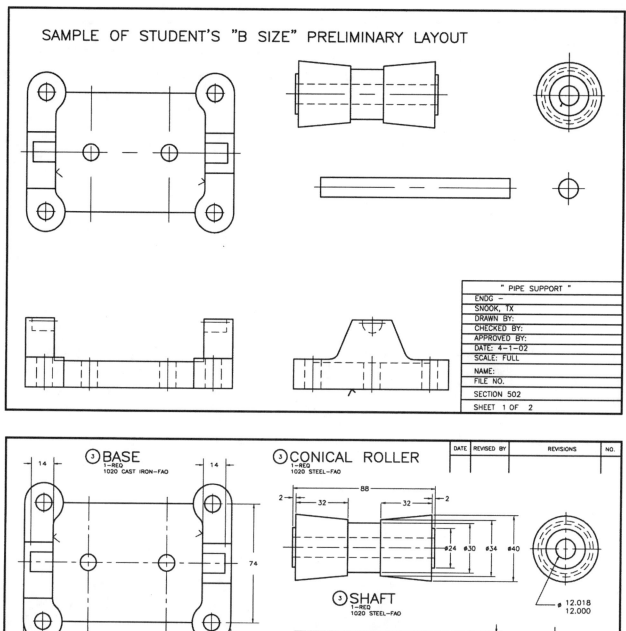
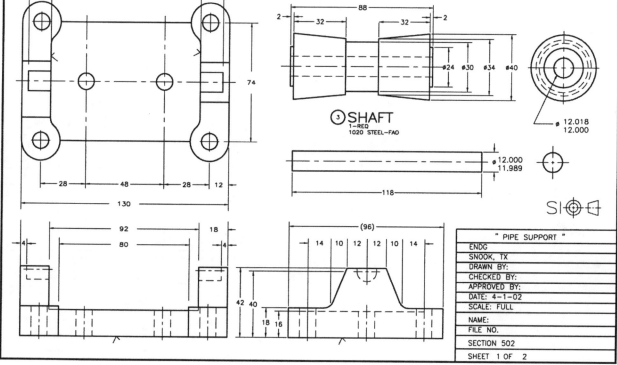

Figure 8.2 Paper is cheap. Plan your page layout to include space for dimensions. Crowding too many parts on a page makes it difficult to separate views and hard to find dimensions.

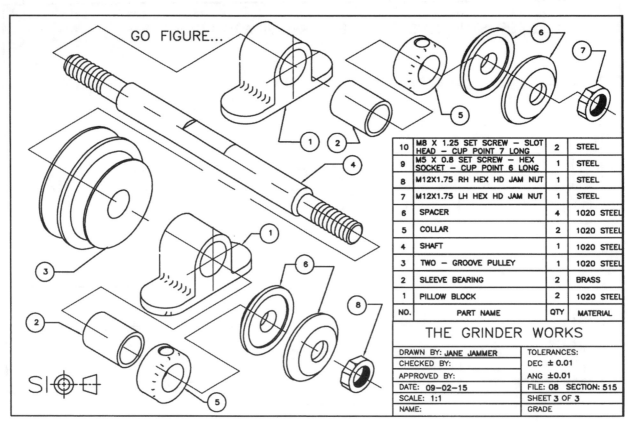

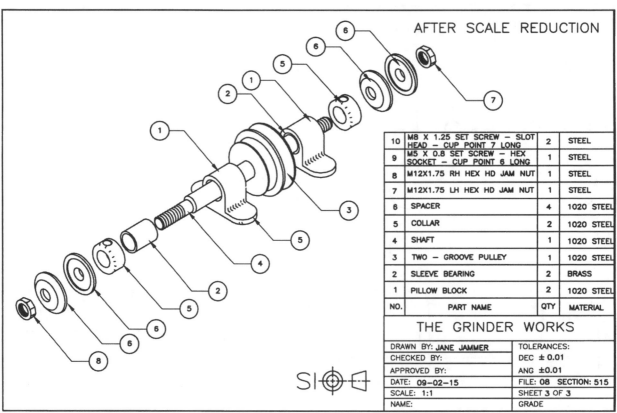

Figure 8.3 Reduce scale or use a larger page as required to present a neat and orderly assembly view. Align balloons in linear patterns when possible.

Working Drawings 149

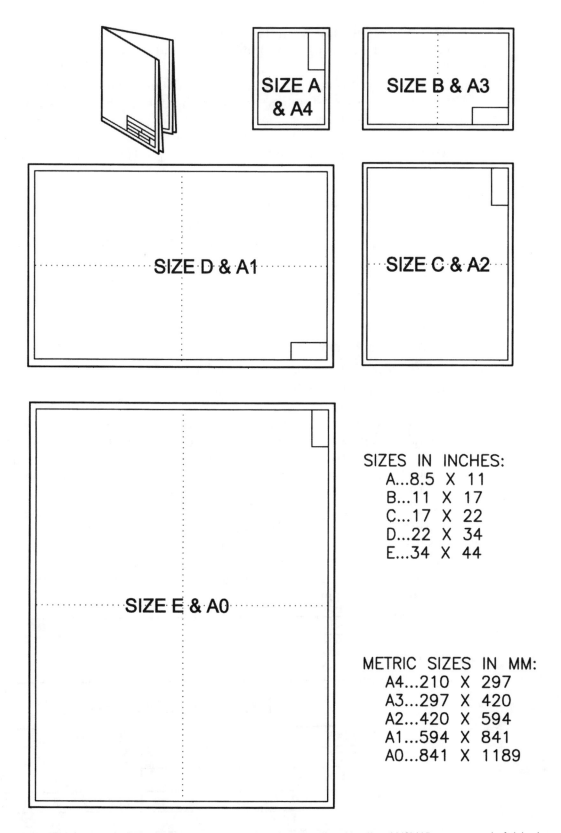

Figure 8.4 Sizes of drawing sheets have been standardized by the ANSI. When properly folded they reduce down to the base size of 8.5" x 11".

be networked worldwide. In large companies with extensive networking, it is just as easy to send a drawing to print out in Scotland or Japan, as it is to send it down the hall.

The main consideration for duplicating drawings should be clear communication. Even the finest laser printers are limited if your original is drastically shrunk down and the text and lines become too small to interpret. That's why the pre-CAD layout sketch is so important. This sketch sets the pattern for scale, text height and line weights. It can be void of details and text since it is only used for planning.

Title Blocks

Each sheet in a set of working drawings must be complete with a title block that is usually on the bottom edge of the drawing and generally in the bottom right hand corner of large sheets. When the sheet is folded down to "A" size, it should always have the title block on the outside of the drawing. **(Figure 8.4)** A complete title block contains the basics of:

1. WHO (drew it)
2. WHAT (is the device, the scale, the sheet tolerance)
3. WHERE (company name and location)
4. WHEN (date it was drawn)

Information for sheet numbers, approvals, part numbers, drawing scales, and general tolerances should also be included. A sample title block is shown in **Figure 8.5**. Most CAD software comes complete with template sheets, borders, and title blocks. AutoCAD© for example, has over 66 choices to match the designers needs of ANSI, ISO, DIN and even some German or Japanese formats. Creating template files is very easy and your company will probably already have custom templates complete with critical information, company name, and logo. Another advantage of using template files is that they can be preloaded with styles of lines and text so the entire company is uniform in its drawing format. In either CAD or hand-filled title blocks, the drawing number should be the dominant feature. It should be twice as tall and twice as bold as the other information.

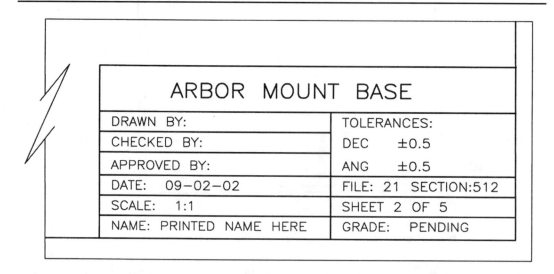

Figure 8.5 The title block should always be in the lower right hand corner. The "drawn by" and "checked by" boxes are for signatures to validate the design.

If the units are metric, an "**SI**" symbol should be placed next to the title block. This symbol indicates the projection system used and that the units used on the sheet are in millimeters (see figure 4.38). Also, it is getting to be common practice to place a scale bar along the bottom edge of the sheet. This aids users who may be working from documents that are printed larger or smaller than the original.

Parts Lists and Revision History

Drawings of assemblies should have a **parts list** placed above the title block. This list should include the part number, name, material and quantity required to build a single unit. **(Figure 8.6)** It may include additional information about vendors or special information like the weight of each part.

When parts lists are drawn by hand, it is traditional to start at the bottom of the list with part number one (No.1) and build your way up to the last part number. (Refer to the previous **Figure 8.6**.) By using this numbering sequence, it is very easy to add additional parts by simply drawing a new space and leaving the other numbers alone. If they were numbered in descending order and nested against the title block, all would have to be erased and reprinted just to add a single part. This is not quite so important with CAD drawing since blocks of texts and lines can be easily moved around and edited.

The revision history normally appears at the upper right corner of the page directly over the title block. This is a critical listing of when the drawing was last revised so persons referring to the sheet can tell if they are using the most current revision. It should include the date, nature of the revision, and the name of who performed the revision. In a well-managed CAD document control system, the revisions to the database are instantaneous worldwide, as users open files for reference. It is important that only the latest version is available for printing. Also, some of the better software has an "engineer's notebook" built in which can be opened by simply clicking on an ICON on or near the part. This tells the revision history or gives additional information about the part like "supersedes parts 362L and 362K" or "use dry lubricant when assembling."

Individual pages should contain complete details, views, dimensions and material notes required to make the part. More than one part can be included on a page as long as it is not crowded. Vendor items like bolts, nuts and washers need to be completely specified but are not drawn on the detail pages. **(Figure 8.7)** They need to be drawn only on the assembly view and even then, they do not require complete details as they will not be manufactured for the device, but rather bought from others.

Assembly Views

As the name implies, **assembly views** show how all of the individual parts and subassemblies are joined to make the device. These views need to have each part labeled with its part number attached to it with a leader. The number at the end of the leader should be circled with a circle diameter approximately 3–4 text heights. The part number should be consistent with the parts list and the part names used on other pages. The text height for these numbers remains .125" so the "balloon" diameter should be between ³/₈" and ¹/₂". Balloons and leaders should never cross each other and should be arranged in a neat and orderly manner.

Assembly views may be fully exploded, partially exploded, or nested together. They may be orthographic or pictorial and may include section views that reveal hidden parts. **(Figures 8.8–8.10)** The important thing is to neatly show a logical

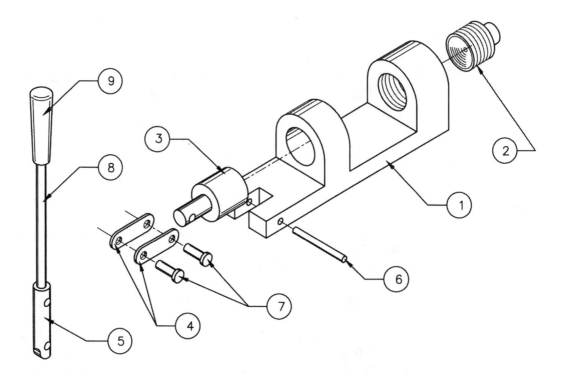

NUTCRACKER PARTS LIST				
9	PUSH KNOB	PLASTIC	1	McMASTER
8	HANDLE SHAFT	STL	1	OUR SHOP
7	RIVET	STL	2	McMASTER
6	AXLE ROD	STL	1	VENDOR
5	ROCKER BASE	STL	1	OUR SHOP
4	RAM LINK	STL	2	TEX STAMP
3	CRUSHER RAM	STL	1	OUR SHOP
2	REVERSIBLE ANVIL	STL	1	OUR SHOP
1	BASE	ALUMINUM	1	U.S. CASTING
NO.	PART NAME	MATERIAL	REQUIRED	SOURCE

Figure 8.6 A complete parts list (often called a bill of material "BOM") is shown with the numbers starting at the bottom as required.

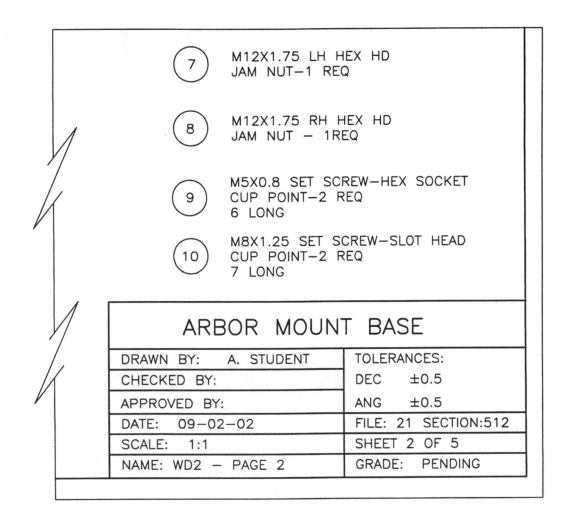

Figure 8.7 Common vendor items are not drawn, but are listed as separate parts where space allows on the detail sheets.

arrangement of how all of the parts are assembled. The views often become title pages and most often precede the other pages of detailed drawings. It is best if the parts list or bill of materials "BOM" appears on the same page but large assemblies may require them to appear on following pages.

Parts shown on the assembly page do not have to include full details, and dimensions, hidden lines and centerlines are usually omitted. The scale of the assembly view may have to be smaller than the scale used in detailed views in order to fit all of the parts on a single page. It's always a good idea to begin the set of drawings with some idea of how your assembly view is going to be presented. It is easy to trace and transfer parts from detailed pages onto the assembly view if they are drawn in the right format from the start. With CAD it is especially easy to cut and paste, or drag and drop parts from other pages. An example of this would be like the part shown in **Figure 8.11**. It is only a two-view drawing but an isometric pictorial was also added for clarity. Later, the isometric part was copied onto a pictorial assembly view for great impact with very little extra effort.

Once the set of drawings is complete, they need to be checked for accuracy of drawing details, format, complete dimensions, and specifications. They are approved with a signature and are forwarded to the design or manufacturing engineer for approval. Only after the major partners in the design have approved the plans can they be released for production. Working drawings are legal documents that may become the center of discussions on the interpretation of production details or in a court of law for patent infringement or product liability. Be sure to "dot every i and cross every t" to make certain the set is complete and accurately reflects the design intent. A little extra effort at this level can prevent much grief in the future.

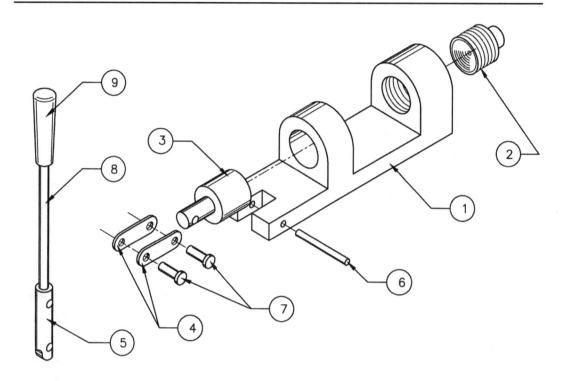

NUTCRACKER PARTS LIST				
9	PUSH KNOB	PLASTIC	1	McMASTER
8	HANDLE SHAFT	STL	1	OUR SHOP
7	RIVET	STL	2	McMASTER
6	AXLE ROD	STL	1	VENDOR
5	ROCKER BASE	STL	1	OUR SHOP
4	RAM LINK	STL	2	TEX STAMP
3	CRUSHER RAM	STL	1	OUR SHOP
2	REVERSIBLE ANVIL	STL	1	OUR SHOP
1	BASE	ALUMINUM	1	U.S. CASTING
NO.	PART NAME	MATERIAL	REQUIRED	SOURCE

Figure 8.8 A complete parts list (often called a bill of material "BOM") is shown with the numbers starting at the bottom as required.

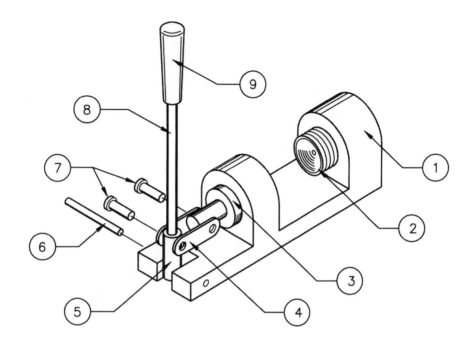

NUT CRACKER PARTS LIST				
9	PUSH KNOB	PLASTIC	1	McMASTER
8	HANDLE SHAFT	STL	1	OUR SHOP
7	RIVET	STL	2	McMASTER
6	AXLE ROD	STL	1	VENDOR
5	ROCKER BASE	STL	1	OUR SHOP
4	RAM LINK	STL	2	TEX STAMP
3	CRUSHER RAM	STL	1	OUR SHOP
2	REVERSIBLE ANVIL	STL	1	OUR SHOP
1	BASE	ALUMINUM	1	U.S. CASTING
NO.	PART NAME	MATERIAL	REQUIRED	SOURCE

Figure 8.9 This "**partially exploded**" assembly view uses an isometric pictorial to show how the mating parts fit together. Always arrange the leaders and balloons into an attractive, well organized, frame around the parts.

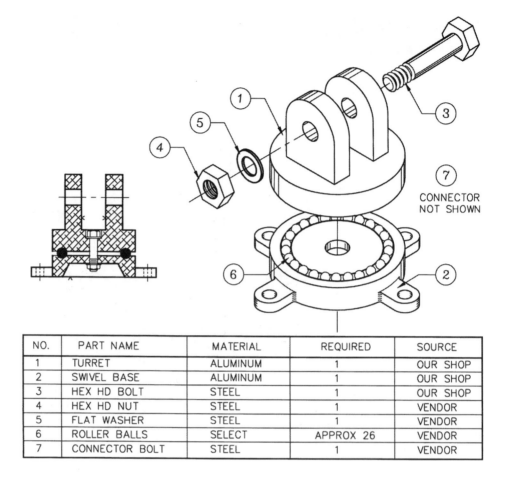

NO.	PART NAME	MATERIAL	REQUIRED	SOURCE
1	TURRET	ALUMINUM	1	OUR SHOP
2	SWIVEL BASE	ALUMINUM	1	OUR SHOP
3	HEX HD BOLT	STEEL	1	OUR SHOP
4	HEX HD NUT	STEEL	1	VENDOR
5	FLAT WASHER	STEEL	1	VENDOR
6	ROLLER BALLS	SELECT	APPROX 26	VENDOR
7	CONNECTOR BOLT	STEEL	1	VENDOR

Figure 8.10 This "**fully exploded**" assembly view uses an isometric pictorial to show how the mating parts fit together. A full section is also shown as an assembly for clarification.

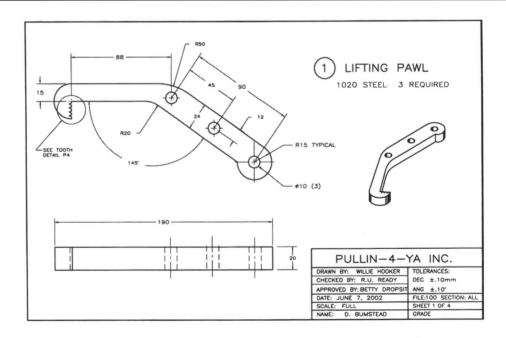

Figure 8.11 A well spaced layout by student with an isometric view included for clarity. Isometric will be copied onto assembly view later. Size A3 sheets shown at a drastically reduced scale.

CHAPTER 8—Practice Quiz: Working Drawings

_____ 1. What is supposed to be the dominant feature in a title block?
A) Name of the part   B) Name of the designer   C) Drawing number   D) Number of parts   E) None of the above

_____ 2. The revision history is normally recorded in the _____ of the page.
A) upper left hand corner   B) lower left hand corner   C) upper right hand corner
D) lower right hand corner   E) none of the above

_____ 3. When creating a hand drawn parts list, it is traditional to list the first part (number 1) at the _____ of the list.
A) top   B) right   C) bottom   D) any of these   E) none of these

_____ 4. Items such as bolts and nuts need to be completely specified and drawn on the detail pages.
A) True   B) False

_____ 5. In assembly views, each part is labeled with its part number attached to it with a _____ .
A) header   B) leader   C) footer   D) title   E) none of these

_____ 6. When selecting paper for the ANSI system, the 34" x 44" is known as Size _____ .
A) A   B) C   C) E   D) F   E) none of these

_____ 7. The main consideration for duplicating drawings should be for _____ .
A) color   B) multiple copies   C) clear communication   D) cost of copies   E) none of these

_____ 8. Parts shown on an assembly page must show full details and often include both hidden and centerlines.
A) True   B) False

_____ 9. At what stage are patent drawings often generated?
A) Layout and analysis   B) Detail and specifications   C) Assembly drawings
D) None of these

_____ 10. List the "4 Ws" basics to be included in a complete title block.
_____ , _____ , _____ , _____

_____ 11. What is the diameter of the balloons used on assembly drawings?
A) 1–2 text heights   B) 3–4 text heights   C) Varies with the scale of the part
D) .125"   E) None of these

_____ 12. The units implied by a scale listed as 1 = 2 are _____ .
A) millimeters   B) centimeters   C) inches   D) either millimeters or centimeters
E) none of these

_____ 13. The SI symbol should be placed on a drawing _____ .
A) in the upper left hand corner   B) next to the title block   C) next to the revisions
D) in the lower left hand corner   E) none of these

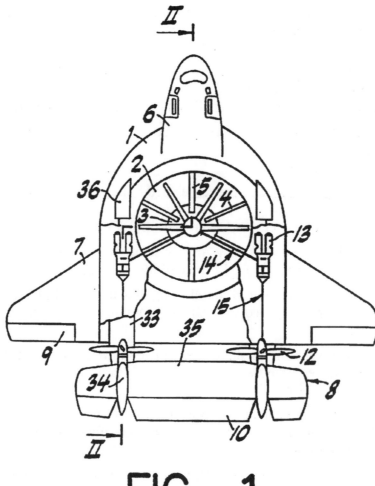

FIG. 1

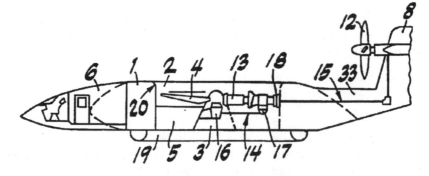

FIG. 2

Engineering Design

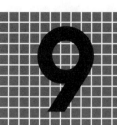

Few, if any, will dispute that design is the single most important aspect of the engineering profession. Design is what engineering is all about. Design is divided into two main categories:

1. Aesthetic—artistic, sometimes abstract
2. Functional—scientific, utilitarian

Engineering design is mainly thought of as functional, but can also be aesthetically pleasing. For example, a motorcycle or automobile may be beautiful, but still has to fulfill its function of moving people from place to place. **(Figure 9.1)** Another difference between aesthetic and functional design is that engineering designs most often are required to solve problems like a broken part, a failing structure, or a modification to upgrade existing parts. These almost always have the limits of time and money attached. In other words, engineers have to complete their designs on time and within budget. Sometimes artists can be commissioned in a similar fashion but they can also just create art for art's sake. Appreciation of both types of design is beneficial to becoming an effective engineer.

The Design Process

Design is often closely linked to creativity and invention. We get our ideas from many sources including books, nature, television, and the World Wide Web. Designers try to be efficient and not "reinvent the wheel" with every new design challenge. They do their homework by analyzing similar designs (reverse engineering), searching through patents, and reviewing technical papers on the same area. For today's design students, the greatest resource of all times is at your fingertips—the World Wide Web. From your computer, you can review last week's technology breakthroughs at "yahoo.com/headlines" or "techweb.com." For deeper immersion, you can review

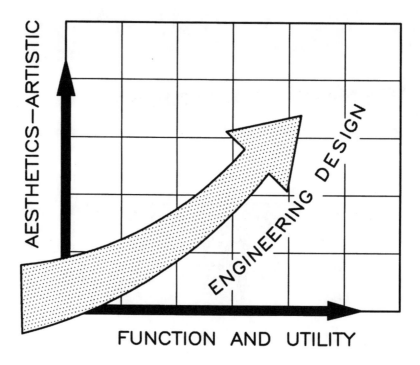

Figure 9.1 Engineering designers should have an appreciation for both the functionality and aesthetics of their designs.

over 6 million patents and their drawings at the United States Patent and Trademark website "USPTO.gov." By seeing what has already been designed, it will give you some insight, if not amazement, for what has been done before us. For pure inspiration, check out some of these great American inventors' life stories at "inventorsmuseum.com" or "mit.edu/invent."

1. Eli Whitney
2. Benjamin Franklin
3. Elijah McCoy (The Real McCoy)
4. Orville and Wilbur Wright
5. Abe Lincoln (only U.S. president with a patent)
6. Charles Goodyear
7. Thomas Edison (he has over 1000 patents)
8. Henry Ford

As in most Web excursions, you will undoubtedly be pulled into more and more links, so kick back and enjoy. As you discover new links of design and invention interests, be sure to share them with your teacher and classmates.

Steps of Engineering Design

While the jury still seems to be split as to whether creativity can be learned or not, there are some well established steps that define the engineering design process. Each author puts the steps in their own words but basically they all seem to ask 3 questions:

1. What's the problem or need? **(Problem)**
2. What are some possible solutions? **(Possibilities)**
3. What is the best choice for the product? **(Products)**

Let's take a quick look at how some other authors describe the design process:

James H. Earle—Identify, Ideate, Refine, Analyze, Decide, Implement

Gary R. Bertoline and Eric Weibe—Ideation, Refinement, Implementation

Giesecke, Mitchell, Spencer, Hill, et.al—Identify Need, Conceptualize, Compromise Solutions, Analyze Models or Prototypes, and Produce Working Drawings

Louis G. Lamit, Kathleen L. Kitto—Problem I.D., Conceptualization, Evaluation, Analysis Decision, Implementation, Production, Marketing

As you can tell, most of these variations say the same thing using their own styles. So, however you choose to describe the process, it still breaks down to the 3P's of: What is the **problem** or need? What are some **possible** solutions? What is the best choice for the **product**?

Design Teams

Before delving too far into some basic design concepts, we need to discuss the role of design teams. Industries discovered the effectiveness of design teams many years ago and strongly embrace the concept. Their teams are comprised of diverse groups of stakeholders who represent engineering, manufacturing, marketing, accounting, legal, etc., to find solutions to design problems or ideas to create new products. The teams meet, analyze the problems, and brainstorm for possible solutions. They make compromises as needed, conduct tests and recommend a final solution.

In this course, you should be provided with the opportunity to work with 3 or 4 other students on a design team. This will prepare you for teaming in upper level classes later and for your work after graduation.

Effective teams have well defined roles for each member and rotate these duties at regular intervals to ensure each member can gain experience at each task. Team members should begin by creating a team "code of conduct" crafted with group input. This defines the behavior expected of each member and is signed and copied so that each person has a copy. Concepts like rotating roles, frequency of meetings, tardiness or absence from team meetings are just a few of the items to be addressed in the code of conduct. Typical roles include coordinator, recorder, timer and gate keeper (to keep the team on-task).

Past experience has shown that students can develop lasting ties and may continue to work together throughout their college careers. (This is not really a new idea; in the early 60's, we called it "cooperate and graduate.") The experience gained by working with a diversity of team members will also enlighten you to issues of our society that you may not yet understand. Diversity is much more than just race or sex; it includes, but is not limited to national origin, speech, sexual orientation, age, intelligence, and rural or urban life experiences. You may get to pick your own teammates or your instructor may assign them. In either case, you would not want to be teamed with persons just like yourself and miss out on divergent ideas and talents the others have to contribute.

Logical Steps from Design to Solution

The individual steps within the 3P's that your team will follow for its design challenge are:

1. **Identify the Problem**—Your team should meet to discuss the design challenge to fix a problem or create a new product. Make sure everyone has a clear understanding of the function, market potential, special requirements, and patent potential of the project.

2. **Explore Possibilities**—Conduct some formal, timed brainstorming sessions where the emphasis is on the quantity of ideas generated rather than the quality. Some great ideas are sometimes planted as seeds by an off-the-cuff, silly remark. Have all ideas recorded by one member on a large pad or white board so everyone can see them. When the ideas stop flowing, get copies of the results for each member. Remember that judgment is reserved and criticism should be avoided during any brainstorming session. Save that for the next phase.

3. **Select Top Three**—Examine and critique each idea to see which ones are the most feasible. Narrow the field of possibilities to the top 3 or 4 ideas. During these critiques, be sure to focus on the issue and not the issuer. Keep personalities out of the decision process and focus on the data. Work as if you had your life savings at stake.

4. **Refine and Test**—Make sketches, models, or prototypes of the top designs and conduct a more critical analysis of their merits and limitations. This might involve mathematical simulation, finite element analysis (FEA), or even destructive testing to see which design emerges as best.

5. **Combine and Improve**—Use results from step 4 to refine and improve the final design. It is possible to combine the best traits of several designs into the best single design. Produce working drawings so that a scale or full size prototype can be built and tested. Final costs can be calculated from these drawings. Conduct market potential surveys using the models.

6. **Verify the Design**—Verify the strength and function for the design. These are of major importance since a product that breaks or is unable to function would not sell. Estimate the production costs and methods.

7. **Write a Summary**—Write a conclusion statement that summarizes your team efforts and final recommendations. Do not think the process is a failure if the conclusion is to not go forward with the design. The team did its job and no money was wasted on production, warehousing, shipping, or advertisement. It is better to turn the idea down now than to loose thousands or millions of dollars later. The design cycle is now finished and all the team has to do is decide how, when, or if to go into production, and what volume to produce for the initial run.

An example of an abandoned design occurred with a student design team that wanted to create a "Shower Caddy/Stool." They designed a very attractive, strong, non-corrosive stainless steel stool to sit on in the shower. It had built in compartments for shampoo, conditioner, razor, washcloths, etc. When they did their economic analysis, they projected it would wholesale for around $40.00 (which included a 25% profit). During the final report stage, a team member discovered a very similar stool at a local discount store that sold for around $15.00 retail. It was made entirely of non-corrosive plastic. They decided against going into production, although there might have been a high-end market from hospitals, hotels, and restaurant kitchens because of the sanitary attributes of stainless steel. Only market tests could tell and that would require actual stools to test, so the project was a wash. Millionaire status will just have to wait.

The process is pretty straight forward except that it could be impacted by several unforeseen variables like the economy, laws, political climate, environmental concerns, and labor problems. In short, use teamwork; don't just work on a team. Four heads should certainly be better than one and can definitely do a lot more "data mining." During any phase of the design process, if the results are not positive at the end of a phase, you may be required to go back to the previous step and try again. This is referred to as "looping" and the potential to loop back as in an "if/or" gate continues throughout the process. In fact, if you later decide to modify the final design, you could continue this looping forever.

Patent Resources

Most important of all, be sure to do a thorough analysis and patent search early in the process so as not to be surprised later. Document each step in an engineer's log and keep other potential ideas in an idea log that you always carry. If you think your idea is deserving of a United States Patent, first check out the process on "USPTO.com" to determine the steps you must complete. If at all possible, consult with a patent lawyer about the process and the potential for your design. Their experience can save you a lot of time and money in the long run and set up the claim so it can be easily defended at a later date. If your discovery was made at work as part of your regular job, then your employer has all of the rights to your idea. Of course, some negotiations are possible to grant you a share of future income. You will need to provide early, witnessed sketches and other verification of when you first came up with the idea. The story goes that another man walked in to file a patent for the telephone about one week after Alexander G. Bell filed his patent, so keep good records and work fast.

The patent process is very strict, very expensive and takes 1–2 years to complete. Patent drawings are required to be produced with instruments (no freehand drawings)

> The Constitution of the United States of America
>
> Article 1, Section 8, Clause 8
>
> The Congress shall have the power...
>
> to promote the progress of science and useful arts by securing for limited times to authors and inventors the exclusive right to their respective writings and discoveries.

164   Chapter 9

on a sheet of white paper that is 10.0" x 15.0" with 1.0" borders on all sides. Work must be done in India ink or a medium that makes excellent copies. The drawings are pictorial or orthographic in nature and include figure numbers and part names. Many other rules apply so check on the Web for the very latest updates. Oddly enough, dimensions are omitted and no proof is required that the design even works in order to be patented. See examples in this chapter of recent patents issued for 2 flying cars and a flying saucer. Actually, the Russian design for a flying car looks pretty good. You may even see them streaking over your local freeway soon.

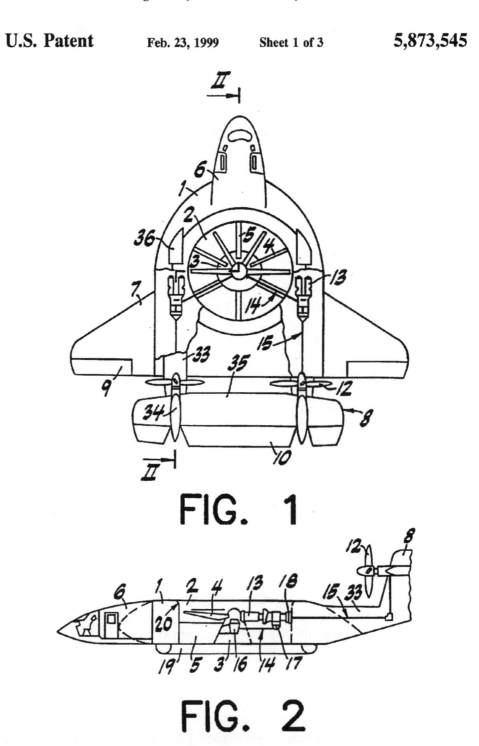

U.S. Patent   Feb. 23, 1999   Sheet 1 of 3   **5,873,545**

FIG. 1

FIG. 2

US006224012B1

(12) United States Patent
Wooley

(10) Patent No.: US 6,224,012 B1
(45) Date of Patent: May 1, 2001

(54) **ROADABLE AIRCRAFT COMBINED VEHICLE FOR PRACTICAL USE**

(76) Inventor: **Donald H. Wooley**, 252 Las Miradas Dr., Los Gatos, CA (US) 95032-7687

(*) Notice: Subject to any disclaimer, the term of this patent is extended or adjusted under 35 U.S.C. 154(b) by 0 days.

(21) Appl. No.: **09/227,286**

(22) Filed: **Dec. 30, 1998**

(51) Int. Cl.[7] B64C 37/00
(52) U.S. Cl. 244/2; 244/49; 244/87; 244/56; 244/121; 244/130
(58) Field of Search .... 244/2, 46, 49, 244/50, 120, 87, 56, 66, 45 R, 55, 90 R, 109, 121, 110, 130

(56) **References Cited**

U.S. PATENT DOCUMENTS

2,384,296	* 9/1945	Gluhareff	244/66
2,505,652	* 4/1950	Schweitzer et al.	244/110 G
3,012,737	* 12/1961	Dodd	244/2
4,401,338	* 8/1983	Caldwell	244/130
4,690,352	* 9/1987	Abdenour et al.	244/130
5,201,478	* 4/1993	Wooley	244/2
5,435,502	* 7/1995	Wernicke	244/2
5,597,137	* 1/1997	Skoglun	244/66
5,984,228	* 11/1999	Pham	244/2

* cited by examiner

Primary Examiner—Galen L. Barefoot

(57) **ABSTRACT**

A vehicle that combines the freedom and swiftness of flight with the utility of surface travel. A simple, light weight, safe and stable conveyance with positive yaw-roll coupling, special safety features, no fold wing stowage and a unique single power source. It can be built using common tools and techniques with readily available materials at a cost competitive with the family automobile.

14 Claims, 8 Drawing Sheets

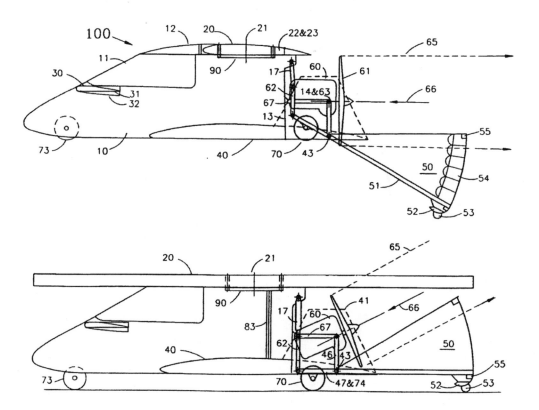

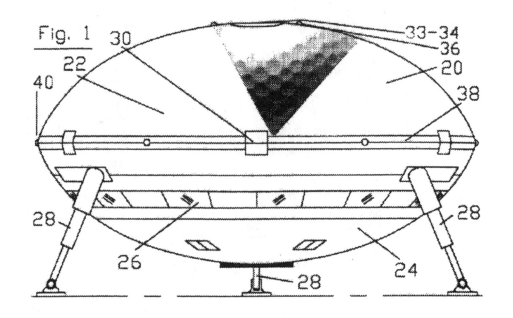
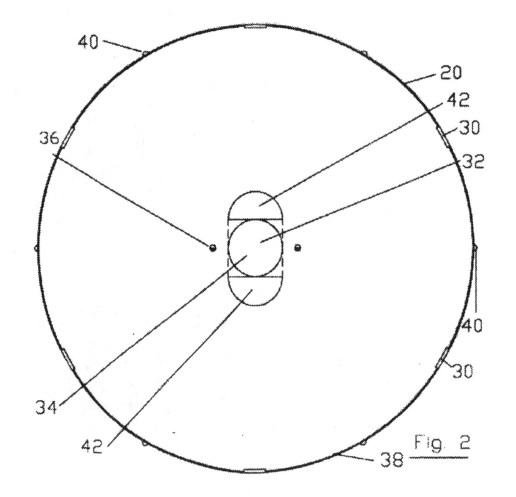

Sample Design Report

The design process may or may not end with a patent. Sometimes it is more expeditious to go into production without a patent just to beat the competition to the market place. By following the steps described earlier in this chapter, you will have a better understanding of how to estimate the market potential, plan your production and possibly earn lots of money. The following section provides a look at a design report of a typical student design project to create an "Ant Proof Dog Food Bowl." This should provide a reasonable model to follow because the same steps used here would work for any design.

Semester Design Project
Spring 2002
Engineering Design Graphics

**PUPPY LOVE PRODUCTIONS
Presents**

THE ANT PROOF DOG FOOD BOWL

The Puppy Love Production Team:

1. Maria Hernandez _____

2. Richard Dakota _____

3. Sue "Peg" Lee _____

4. Jerome Jackson _____

Problem Identification

1. **Project Title**

    Ant Proof Dog Dish

2. **Identify the Problem**

    Ants crawl into pet's food bowl, bite pets, and ruin food.

3. **Minimum Requirements**

    A. Cheap—sell for less than $10.00
    B. Easy to assemble
    C. Little or no maintenance
    D. Durable—pet proof, corrosion resistant, crush resistant
    E. Safe for pets and children
    F. Size—must hold minimum of two 16 oz cans of dog food
    G. Dogs must like it

4. **Discovery Process**

    A. What is average size of dog bowl?
       Inquire at retail stores
    B. How can ants be prevented from climbing into food?
       Brainstorming and experimentation . . .
    C. What are conventional dishes made of?
       Inquire at retail stores.
    D. How much does average dog eat each day? Serving size?
    E. How can we keep it sanitary?
    F. Ant climbing dynamics—check with Entomology Department.
       Do environmentalists protect ants?
    G. What is the range of dog sizes we want to accommodate?
    H. What do dogs prefer? Conduct tests.

5. **Market Consideration Process**

    A. Who can make product at the cheapest price?
       Inquire at manufacturing plants
    B. Who would sell product:
       Check yellow pages for retail stores
    C. Are there any other ant proof dishes on the market?
       Inquire at retail stores, catalogs, patent library.
    D. What is the sales price of conventional dishes

Possibilities for Solution

1. **Brainstorming Process**
    A. Double Decker bowl
    B. Poison rim

C. Electric current at base
D. Slippery surface
E. Electric rim
F. Fire underneath
G. Hanging bowl
H. Angled sides
I. Lip above ground
J. Pet anteater
K. Bowls on top of each other
L. Bowl stand
M. Electric ant call
N. Strap around bowl
O. Rounded lip
P. Ant motel
Q. Zapper light bottom
R. Put bowl on spikes
S. Food dispenser
T. Built in ant motel
U. Ant mine field
V. Put over another bowl of water

Compare, Refine, and Compromise Process

1. **Description of Designs** (see attached sketches)
    A. Rounded lip design
        1. Conventional dish design with curved under lip
        2. One piece molded plastic design
        3. Ants cannot climb around the inverted lip
    B. Water Moat Design
        1. Conventional dish on stand surrounded by water moat
        2. Two part construction of molded plastic
        3. Ants trying to get into bowl get trapped in moat and drown—can double as a water dish for pet.
    C. Ant Motel Design
        1. Conventional dish design with holes in side which lead to poison ant motel
        2. One motel in center of dish at bottom
        3. Motel poison must be replaced occasionally
        4. Ants enter motel instead of dish and die

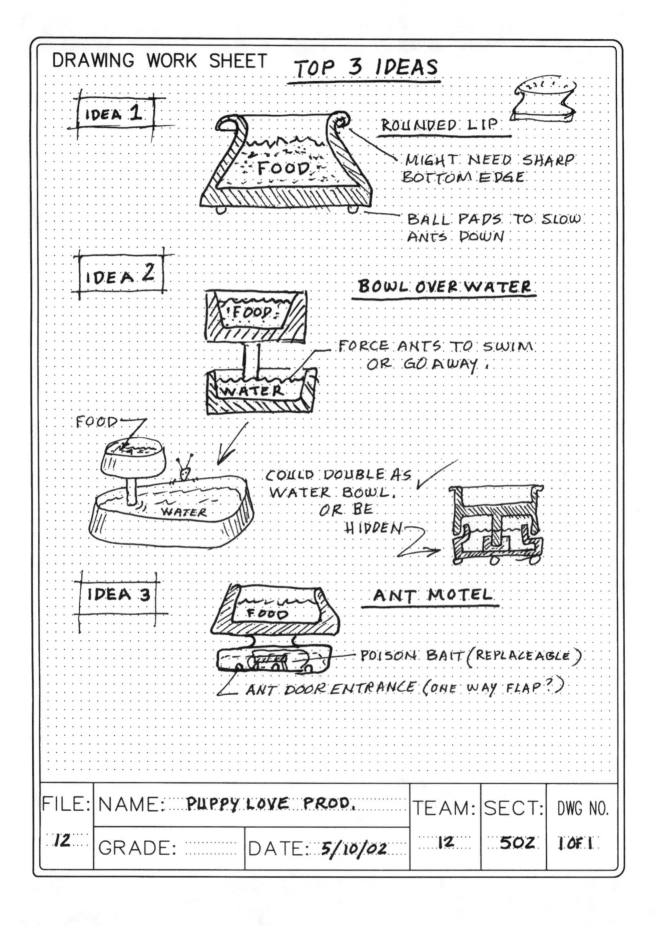

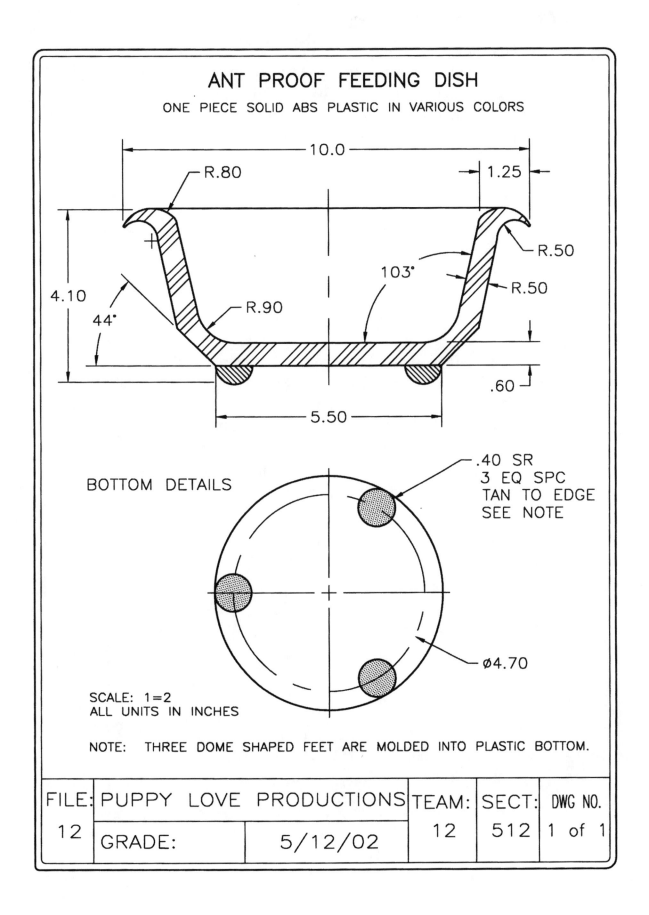

Critical Analysis of Best Design

1. **Proof of Functionality**
   A. Curved lip prevents ants from being able to climb into bowl
   B. Simple design, no moving parts
   C. Solid construction so it is stable and durable

2. **Human and Pet Engineering**
   A. No assembly required
   B. No maintenance required
   C. Safe for pet and children

3. **Marketing Considerations**
   A. Keep price under $10.00
   B. Can be sold in conjunction with dog food and pet supplies
   C. Need to estimate how many conventional dishes are sold yearly
   D. Should make several sizes in future—large dog, small dog, cat, etc.
   E. Worldwide potential

4. **Physical Description of the Bowl** (see attached drawings)
   A. Solid one piece construction in shape of conventional dog dish
   B. Curved overhanging top lip to prevent ants from climbing into food
   C. See scale drawings attached

5. **Verify Strength and Durability**
   A. Solid plastic design is same used in conventional dog dishes, which are capable of withstanding the chewing of the largest dogs. Also, the same type of material is used for making football helmets, so it should be durable. Drop and abrasion tests were conducted in our lab and results were very favorable. The technical report is available upon request.

6. **Production Process**
   A. Use conventional plastic injection molding procedures to create one piece molded plastic dish as specified in final drawing.

7. **Economic Analysis**
   A. Cost per bowl
   1. Plastic $ .50
   2. Labor $ .75
   3. Packaging $ .30
   4. Profit $3.00
   Wholesale Price $4.55
   Suggested Retail Price $6.50

8. Summary and Conclusions

The Ant Proof Pet Bowl can be manufactured and distributed for $4.55 per bowl and would be sold to retailers in cartons containing 10 bowls. We would expect to sell 10,000 bowls during our first year for a projected profit of $30,000. We will seek a start-up loan of $50,000 for tooling, materials, distribution, and labor to produce the first run. We anticipate that future production runs will be more profitable since the molds can be reused.

This looks like a profitable venture, and we plan to meet with representatives of the local Small Business Administration Office to explore grant and loan possibilities. We also plan to meet with representatives from a patent lawyers office to see about protecting our design rights with a U.S. Patent. All things considered, we see this as a good investment and hope to begin production within 90 days. We will let you know when stock is available since we are confident you should become an investor in Puppy Love Productions.

CHAPTER 9—Practice Quiz: Engineering Design

_____ 1. For today's engineering student, one of the greatest resources of all times for design ideas is the campus library.
A) True   B) False

_____ 2. All applications for patents must be accompanied by a set of drawings showing figure numbers and part names, but dimensions are not required.
A) True   B) False

_____ 3. Before final patent approval, the designer must prove the invention really works.
A) True   B) False

_____ 4. When choosing a team member, you should select someone with similar interests and background as yourself.
A) True   B) False

_____ 5. After a design team has been selected, the members start by defining the _____ .
A) products to be developed   B) roles of each member   C) team code of conduct
D) requirements for a patent   E) none of these

_____ 6. Patent drawings must be submitted on a sheet of paper that is _____ .
A) 8.5" x 11"   B) 11" x 17"   C) 10" x 15"   D) 10" x 12"   E) depends on the invention's size

_____ 7. During brainstorming sessions, only serious ideas should be recorded.
A) True   B) False

_____ 8. When searching for ideas for your product, it is not acceptable to review other's designs and inventions.
A) True   B) False

_____ 9. The single most important aspect of engineering is design.
A) True   B) False

_____ 10. In industry, if your design is rejected for production, then the whole process was a waste of time and money.
A) True   B) False

_____ 11. Sometimes artistic designers have similar constraints for being on time and within budget like engineers.
A) True   B) False

12. Name 3 typical roles of members of a design team.
A) _____   B) _____   C) _____

13. Name 3 items that should be included in the "code of conduct."
A) _____   B) _____   C) _____

14. Name the 2 main categories of design.
A) _____   B) _____

15. What are the 3 established steps (3 P's) that define the engineering design process?
A) _____   B) _____   C) _____

CHAPTER 9—Design Project Ideas

1. Pickup truck bumper toolbox
2. Bed desk
3. Bath tub desk
4. Motorcycle trailer that converts to a boat
5. Tailgate party box/organizer
6. Automatic plant waterer
7. LCD blackboard
8. Improved slacks hanger
9. Expandable flower pot
10. Bicycle umbrella
11. Hall seating for campus buildings
12. Laptop "lap desk/security locker"
13. Storage seat for sporting events
14. Fishing lure shooter/launcher
15. Retriever for hung fishing lures
16. Rain top for cycle riders
17. Retractable motorcycle cover
18. Shower organizer/entertainment center
19. Remote site gate opener
20. Medicine bottle opener
21. Better poop-scooper
22. Fishing stool/tackle box
23. 2-wheel drive mountain bike
24. Automatic glasses cleaner
25. Motorized fingernail clipper
26. Air conditioned motorcycle helmet
27. Organizer for the professor
28. Single hand keyboard
29. Omni directional oscillating fan
30. Cool blanket
31. Mount for hands-free hair dryer
32. Automatic clothes sorter
33. Time locked pantry
34. Underground tunnel system for campus
35. Vacuum tube mail transport for campus
36. Paperback book clamp to hold it open
37. Basketball retriever
38. Scuba gear organizer/cart
39. Slow pitch machine for softball
40. Volleyball setter/server
41. Auto animal feeder
42. Tennis ball retriever
43. Scuba communicators
44. Golf ball locator
45. Better windshield wipers
46. Voice activated lock
47. Ant proof pet feeder
48. Toilet desk
49. Automatic bicycle gear shifter
50. All wheel drive mountain bike
51. Shoes with interchangeable soles
52. Car CD holder
53. Golf tee planting device
54. Bathroom supply organizer
55. Increasing volume alarm clock
56. Automatic 7-day dog feeder
57. Solar heated pool
58. Car wheel rockets
59. Voice translator
60. Campus trolley system
61. Caulk ball for tennis (in/out boundary)
62. Anti-glare computer screen
63. Nails that don't bend
64. Remote control mini blinds
65. Dorm loft kit
66. Left-handed desk
67. Storage compartment bumper
68. Pillow cooler

69. Motorcycle jack
70. Jig saw circle/curve cutting device
71. Mini-blind timer/programmer
72. Sail skateboard attachment
73. Handicap access mechanism for door
74. Page turner for handicapped
75. Solar fan to cool car in sun
76. Child proof cabinet
77. Cross campus skywalk—automatic
78. Fishing pole with automatic caster
79. Loft that doesn't need tools
80. Etch-A-Sketch chalk board
81. Swimming pool cover—automatic
82. Car auto pilot
83. Animal escape from swimming pool
84. Tray to hold food/drinks in stick shift car
85. Device to direct rain water to plants under patio cover
86. Portable wet bar for tailgate parties
87. Briefcase that opens into a bed
88. Way to carry books other than backpack
89. Easier way to make beds
90. Alarm clock to set off lights for wakeup
91. Horse saddle with shock absorber
92. A baseball cap that holds keys/wallet
93. Automatic lifeguard for swimming pool
94. Rapid evacuation device for tall buildings
95. Kid finder device for lost kids
96. Dog collar with built in leash
97. Sports nut all purpose reclining chair
98. Anti-theft bike rack
99. Motorcycle air conditioning system
100. Portable motorcycle garage

T. A. EDISON.
Electric-Lamp.

No. 223,898.   **Patented Jan. 27, 1880.**

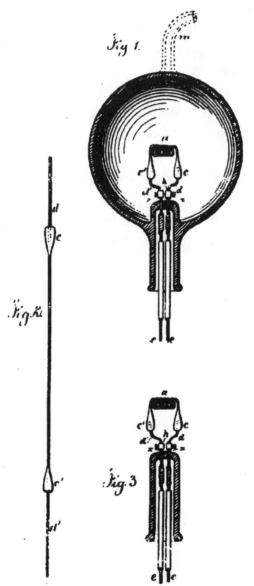

Computer Aided Drafting

Historical Concepts

Computer graphics had its birth in 1963 at MIT when a graduate student named Ivan Sutherland completed his Ph.D. dissertation on interactive computer graphics. Prior to that time, only a computer programmer could get a picture generated which was from combinations of alpha characters like X, O and * printed on long sheets of computer paper. Sometimes, these were used to print out coordinates that were traced with India ink using traditional drafting equipment. The only true computer graphic work centers were so large and expensive that only companies like General Motors or Boeing could afford to purchase and support them. The input of data came from decks of punched computer cards and data was stored on reels of magnetic tapes that were about 12.0" in diameter and an inch thick. Generally, the process of producing a simple drawing took days if not weeks and there would only be one printer which ran 24 hours a day to meet the demand. Much of the time involved was waiting for printouts. With the advent and worldwide acceptance of personal computers, the computer graphics programs flourished with many startup companies in the chase for market shares. These were still relatively expensive, somewhat large, and slow computers by today's standards. The hard drives were only about 10 megabytes as compared to the 40 gigabytes today. Only black and white monitors were available and printers were dot matrix. Graphics could be sent to plotters, which moved ink pens in a positive or negative "x" direction while the paper was moved simultaneously in the "y" direction. When the plots were finished, they had to set for a few minutes to let the ink dry before handling. Pens did not last long and were expensive to replace. To produce different line sizes, you had to use different pens (i.e. P.7 for thick lines and P.3 for thin lines). Color pens were available but black was used most often.

Common Software Concepts

Costs for both hardware and software have dropped in recent years and computer aided drafting (CAD) software choices have been reduced to a handful like AutoCAD©, AutoCAD Inventor©, Mechanical Desktop©, SolidWorks©, CATIA©, Pro/ENGINEER©, MircoStation© and CADKEY© controlling most of the world market. Each of these software packages have advantages and disadvantages depending on the intended applications like architectural, civil engineering, mechanical engineering or parametric solid modeling. For complete and up to date information, check out their Web sites.

What CAD means to drafting and design could be compared to what the telephone meant to the telegraph. All things considered, CAD is just one more tool for the engineer to use in creating designs. There are many advantages that CAD offers over traditional hand drawings. Among these are:

1. Ease and speed of editing and making corrections, which includes functions such as move, copy, erase, stretch, rotate, array, and scaling.
2. Ability to copy parts from one drawing file to another.
3. Ability to send drawings electronically to anywhere in the world in an instant.
4. Ability to store large drawing files in very small space.
5. Ability for multiple users to access the data from any place in the world.
6. Ability to develop template files that have uniform text styles and linetypes preset for each user.
7. Smaller space required than drafting tables.
8. Portability afforded in the field via laptops.
9. Ability to quickly check for fits or interference of mating parts.
10. A drawing file can generate rapid prototypes.
11. Drawing files can go directly to the production machine making the real part.
12. Consistency of style between multiple users.
13. Ability to insert standard parts from parts libraries and second party vendors.
14. More user friendly for handicapped persons.
15. Accurate to within about .0001" for construction.
16. Much quicker to make prints of your work.
17. User groups and online help are available.
18. Line quality has dark and consistent linetypes.
19. Prints are odorless, dry and faster than Diazo type prints.
20. Retrieval of drawing files from archives is faster and can occur from remote jobsites.
21. Hand eye coordination and artistic talent are not as important.

There are probably many more that you can think of to discuss with your classmates or teacher. On the other side, are there advantages of using pencil and paper over computer? A few possibilities are listed below:

1. No electricity or batteries are required.
2. Cost of workstation about $1/5000$ of the cost of a PC and CAD software.
3. Completely portable—take sketchpad anywhere.
4. Sketching is quick and fun.
5. No new software to learn or upgrades to make.
6. No loss of data because of power failures.
7. Easier to develop your own personal style.

CAD certainly has the most advantages, but among these, probably the most significant trends are toward parametric solid modeling (PSM), computer aided manufacturing (CAM) and rapid prototyping (RP). Before expanding on these, there are a few acronyms you should know:

1. CNC—computer numerical controlled (machine)
2. CAD/CAM—computer aided drafting and manufacturing
3. JIT—just in time
4. CAE—computer-aided engineering
5. CIM—computer-integrated manufacturing
6. FEA—finite-element analysis
7. FEM—finite-element meshing and modeling

Specifications to Solids

The future of engineering design graphics is heading toward the use of parametric solid modeling. In this format, the designer sculpts the shape of the design as a solid model on the screen much the same way an artist would sculpt with clay. The solid model is built with all of the feature sizes and geometry accurately input as "parameters" to control their behavior. From a completed parametric solid model, you can generate dimensioned 2-D layouts, bill of materials and links to spreadsheets. Values can be edited either on the spreadsheet or the solid model and both systems will change instantly.

Before too much longer, engineers may not need to make paper printouts except for use in the field. Even this use may not be necessary much longer as portable laptop computers become more popular.

Rapid Prototypes

The solids created parametrically also contain all of the size and geometric databases to produce rapid prototypes and drive CNC machines. This direct link is both practical and intelligent. Why send a drawing that is precise to .0001" to a worker knowing that he will transfer the dimensions onto the metal with a hand held square and a dull pencil? Send the data instead, to a machine that can maintain the .0001" accuracy without ever seeing a printout. Numerically controlled machines are not new technology but are getting more use as prices of computer components drop. Even now, just as designers send their solid models to a printer or rapid prototype machine, in the future, they will just send it directly to the CNC machine for the production of the real part.

The term rapid prototype is a relative term as compared to making the prototype by hand in a model shop. Some prototype machines take approximately one hour to build 1.0" in height resulting in models taking 12 or more hours to build. Compare this however, to 3 weeks in a model shop and it seems rapid. There are 3 R's that describe the ideal preliminary design prototype—**Rough**, **Rapid** and **Right**. These satisfy the need for preliminary analysis and testing. For customer review, a more finished model will be built which provides the client with the 4 F's of rapid prototyping—**Form**, **Fit**, **Function**, and **Feel**. Three of these can be validated on the monitor, but the Feel is where the payoff lies for clear communication and understanding. Files can be sent electronically to subcontractors who do nothing but create rapid prototypes thus saving the investment cost of owning your own machine.

Among the RP processes in use today, there are the:

1. Laminate Object Machine (LOM) that uses paper to build models.
2. Stereolithography Machine (STL) that uses laser guided beams to fuse polymer to make a plastic model.
3. Robotically Guided Extrusion (RGE) that uses a "glue gun" to make very accurate and very tough plastic models.
4. Selective Laser Sintering (SLS) that uses a laser beam to fuse powered polymer beads together to form the plastic model.
5. 3-D Printer that produces a wax model fairly fast and inexpensively but thin sections are brittle and have to be handled very gently.

Experiments with new RP concepts are ongoing and it is an exciting emerging field. Recent experiments have used water-to-ice and heat-to-bake models. One of the faster model makers (by Zcorp) uses a blend of cornstarch that produces a rough but durable model. Who knows, before long, these will be just as common and fast as copy machines are today.

It even gets better. Voice recognition software was developed several years ago but has been difficult to set up and control for accuracy. It is still evolving and being fine-tuned to a point where fairly complex parts are being created very quickly using this innovation. No keyboard, mouse or pointer are used—just a headset and microphone. What a quantum leap for us and especially for the physically challenged who want to practice engineering. These are exciting times and the best is yet to come.

CHAPTER 10—Practice Quiz: Computer Aided Drafting

____ 1. CAD stands for ____ .
A) Computer Aided Drafting   B) Computer Aided Detailing   C) Computer Aligned Drilling   D) Calculate and Dimension   E) None of these

____ 2. Early computers stored files on ____ .
A) 3" disks   B) 5" disks   C) 12" disks   D) CDs   E) none of these

____ 3. STL is the name of a process for making rapid prototypes.
A) True   B) False

____ 4. Which of the following is *not* a popular engineering computer aided software?
A) MicroCAD   B) CATIA   C) AutoCAD   D) SolidWorks   E) Pro/ENGINEER

____ 5. The process of generating a model on the rapid prototype machines has become very fast and accurate.
A) True   B) False

____ 6. Of the 4 F's of rapid prototyping, which one is considered to payoff the best for clear communication?
A) Form   B) Fit   C) Function   D) Feel   E) None of these

____ 7. Rapid prototyping is considered one of the emerging fields in engineering.
A) True   B) False

____ 8. If modifications are needed for a 2-D layout from a parametric solid model, the corrections have to be made on the solid model and the spreadsheet separately.
A) True   F) False

____ 9. Computer graphics was born in 1983 when Apple Computers were first invented.
A) True   B) False

10. Name 4 advantages of using a CAD program for engineering projects.

    A) _____   B) _____   C) _____

    D) _____

11. Name the 3 R's describing the ideal design prototype.

    1) _____   2) _____   3) _____

12. Name 3 advantages of hand drawing over CAD drawings.

    1) _____   2) _____   3) _____

Appendix A

Table 1. Unified Standard Screw Thread Series

| Sizes | | Basic Major Diameter | Threads Per inch | | | | | | | | | | | Sizes |
| Primary | Secondary | | Series with graded pitches | | | Series with constant pitches | | | | | | | | |
			Coarse UNC	Fine UNF	Extra fine UNEF	4UN	6UN	8UN	12UN	16UN	20UN	28UN	32UN	
0		0.060	–	80	–	–	–	–	–	–	–	–	–	0
	1	0.073	64	72	–	–	–	–	–	–	–	–	–	1
2		0.086	56	64	–	–	–	–	–	–	–	–	–	2
	3	0.099	48	56	–	–	–	–	–	–	–	–	–	3
4		0.112	40	48	–	–	–	–	–	–	–	–	–	4
5		0.125	40	44	–	–	–	–	–	–	–	–	–	5
6		0.138	32	40	–	–	–	–	–	–	–	–	UNC	6
8		0.164	32	36	–	–	–	–	–	–	–	–	UNC	8
10		0.190	24	32	–	–	–	–	–	–	–	–	UNF	10
	12	0.216	24	28	32	–	–	–	–	–	–	UNF	UNEF	12
1/4		0.250	20	28	32	–	–	–	–	–	UNC	UNF	UNEF	1/4
5/16		0.3125	18	24	32	–	–	–	–	–	20	28	UNEF	5/16
3/8		0.375	16	24	32	–	–	–	–	UNC	20	28	UNEF	3/8
7/16		0.4375	14	20	28	–	–	–	–	16	UNF	UNEF	32	7/16
1/2		0.500	13	20	28	–	–	–	–	16	UNF	UNEF	32	1/2
9/16		0.5265	12	18	24	–	–	–	UNC	16	20	28	32	9/16
5/8		0.625	11	18	24	–	–	–	12	16	20	28	32	5/8
	11/16	0.6875	–	–	24	–	–	–	12	16	20	28	32	11/16
3/4		0.750	10	16	20	–	–	–	12	UNF	UNEF	28	32	3/4
	13/16	0.8125	–	–	20	–	–	–	12	16	UNEF	28	32	13/16
7/8		0.875	9	14	20	–	–	–	12	16	UNEF	28	32	7/8
	15/16	0.9375	–	–	20	–	–	–	12	16	UNEF	28	32	15/16
1		1.000	8	12	20	–	–	UNC	UNF	16	UNEF	28	32	1
	1-1/16	1.0625	–	–	18	–	–	8	12	16	20	28	–	1-1/16
1-1/8		1.125	7	12	18	–	–	8	UNF	16	20	28	–	1-1/8
	1-3/16	1.1875	–	–	18	–	–	8	12	16	20	28	–	1-3/16
1-1/4		1.250	7	12	18	–	–	8	UNF	16	20	28	–	1-1/4
	1-5/16	1.3125	–	–	18	–	–	8	12	16	20	28	–	1-5/16
1-3/8		1.375	6	12	18	–	UNC	8	UNF	16	20	28	–	1-3/8
	1-7/16	1.4375	–	–	18	–	6	8	12	16	20	28	–	1-7/16
1-1/2		1.500	6	12	18	–	UNC	8	UNF	16	20	28	–	1-1/2
	1-9/16	1.5625	–	–	18	–	6	8	12	16	20	–	–	1-9/16
1-5/8		1.625	–	–	18	–	6	8	12	16	20	–	–	1-5/8
	1-11/16	1.6875	–	–	18	–	6	8	12	16	20	–	–	1-11/16
1-3/4		1.750	5	–	–	–	6	8	12	16	20	–	–	1-3/4
	1-13/16	1.8125	–	–	–	–	6	8	12	16	20	–	–	1-13/16
1-7/8		1.875	–	–	–	–	6	8	12	16	20	–	–	1-7/8
	1-15/16	1.9375	–	–	–	–	6	8	12	16	20	–	–	1-15/16
2		2.000	4-1/2	–	–	–	6	8	12	16	20	–	–	2
	2-1/8	2.215	–	–	–	–	6	8	12	16	20	–	–	2-1/8
2-1/4		2.250	4-1/2	–	–	–	6	8	12	16	20	–	–	2-1/4
	2-3/8	2.375	–	–	–	–	6	8	12	16	20	–	–	2-3/8
2-1/2		2.500	4	–	–	UNC	6	8	12	16	20	–	–	2-1/2
	2-5/8	2.625	–	–	–	4	6	8	12	16	20	–	–	2-5/8
2-3/4		2.750	4	–	–	UNC	6	8	12	16	20	–	–	2-3/4
	2-7/8	2.875	–	–	–	4	6	8	12	16	20	–	–	2-7/8
3		3.000	4	–	–	UNC	6	8	12	16	20	–	–	3
	3-1/8	3.125	–	–	–	4	6	8	12	16	–	–	–	3-1/8
3-1/4		3.250	4	–	–	UNC	6	8	12	16	–	–	–	3-1/4
	3-3/8	3.375	–	–	–	4	6	8	12	16	–	–	–	3-3/8
3-1/2		3.500	4	–	–	UNC	6	8	12	16	–	–	–	3-1/2
	3-5/8	3.625	–	–	–	4	6	8	12	16	–	–	–	3-5/8
3-3/4		3.750	4	–	–	UNC	6	8	12	16	–	–	–	3-3/4
	3-7/8	3.875	–	–	–	4	6	8	12	16	–	–	–	3-7/8
4		4.000	4	–	–	UNC	6	8	12	16	20	–	–	4
	4-1/8	4.125	–	–	–	4	6	8	12	16	–	–	–	4-1/8
4-1/4		4.250	–	–	–	4	6	8	12	16	–	–	–	4-1/4
	4-3/8	4.375	–	–	–	4	6	8	12	16	–	–	–	4-3/8
4-1/2		4.500	–	–	–	4	6	8	12	16	–	–	–	4-1/2
	4-5/8	4.625	–	–	–	4	6	8	12	16	–	–	–	4-5/8
4-3/4		4.750	–	–	–	4	6	8	12	16	–	–	–	4-3/4
	4-7/8	4.875	–	–	–	4	6	8	12	16	–	–	–	4-7/8
5		5.000	–	–	–	4	6	8	12	16	–	–	–	5
	5-1/8	5.125	–	–	–	4	6	8	12	16	–	–	–	5-1/8
5-1/4		5.250	–	–	–	4	6	8	12	16	–	–	–	5-1/4
	5-3/8	5.375	–	–	–	4	6	8	12	16	–	–	–	5-3/8
5-1/2		5.500	–	–	–	4	6	8	12	16	–	–	–	5-1/2
	5-5/8	5.625	–	–	–	4	6	8	12	16	–	–	–	5-5/8
5-3/4		5.750	–	–	–	4	6	8	12	16	–	–	–	5-3/4
	5-7/8	5.875	–	–	–	4	6	8	12	16	–	–	–	5-7/8
6		6.000	–	–	–	4	6	8	12	16	–	–	–	6

Source: Reprinted courtesy of The American Society of Mechanical Enginneers

Table 2. Thread Sizes and Dimensions

Nominal Size		Diameter				Tap Drill (for 75% thread)			Threads per Inch		Pitch (mm)		Threads per Inch (Approx.)	
		Major		Minor										
Inch	mm	Inch	mm	Inch	mm	Drill	Inch	mm	UNC	UNF	Coarse	Fine	Coarse	Fine
–	M1.4	0.055	1.397	–	–	–	–	–	–	–	0.3	0.2	85	127
0	–	0.060	1.524	0.0438	1.092	3/64	0.0469	1.168	–	80	–	–	–	–
–	M1.6	0.063	1.600	–	–	–	–	1.25	–	–	0.35	0.2	74	127
1	–	0.073	1.854	0.0527	1.320	53	0.0595	1.1499	64	–	–	–	–	–
1	–	0.073	1.854	0.0550	1.397	53	0.0595	1.499	–	72	–	–	–	–
–	M.2	0.079	2.006	–	–	–	–	1.6	–	–	0.4	0.25	64	101
2	–	0.086	2.184	0.0628	1.587	50	0.0700	1.778	56	–	–	–	–	–
2	–	0.086	2.184	0.0657	1.651	50	0.0700	1.778	–	64	–	–	–	–
–	M2.5	0.098	2.489	–	–	–	–	2.05	–	–	0.45	0.35	56	74
3	–	0.099	2.515	0.0719	1.828	47	0.0785	1.981	48	–	–	–	–	–
3	–	0.099	2.515	0.0758	1.905	46	0.0810	2.057	–	58	–	–	–	–
4	–	0.112	2.845	0.0795	2.006	43	0.0890	2.261	40	–	–	–	–	–
4	–	0.112	2.845	0.0849	2.134	42	0.0935	2.380	–	48	–	–	–	–
–	M3	0.118	2.997	–	–	–	–	2.5	–	–	0.5	0.35	51	74
5	–	0.125	3.175	0.0925	2.336	38	0.1015	2.565	40	–	–	–	–	–
5	–	0.125	3.175	0.0955	2.413	37	0.1040	2.641	–	44	–	–	–	–
6	–	0.138	3.505	0.0975	2.464	36	0.1065	2.692	32	–	–	–	–	–
6	–	0.138	3.505	0.1055	2.667	33	0.1130	2.870	–	40	–	–	–	–
–	M4	0.157	3.988	–	–	–	–	3.3	–	–	0.7	0.35	36	51
8	–	0.164	4.166	0.1234	3.124	29	0.1360	3.454	32	–	–	–	–	–
8	–	0.164	4.166	0.1279	3.225	29	0.1360	3.454	–	36	–	–	–	–
10	–	0.190	4.826	0.1359	3.429	26	0.1470	3.733	24	–	–	–	–	–
10	–	0.190	4.826	0.1494	3.785	21	0.1590	4.038	–	32	–	–	–	–
–	M5	0.196	4.978	–	–	–	–	4.2	–	–	0.8	0.5	32	51
12	–	0.216	5.486	0.1619	4.089	16	0.1770	4.496	24	–	–	–	–	–
12	–	0.216	5.486	0.1696	4.293	15	0.1800	4.572	–	28	–	–	–	–
–	M6	0.236	5.994	–	–	–	–	5.0	–	–	1.0	0.75	25	34
1/4	–	0.250	6.350	0.1850	4.699	7	0.2010	5.105	20	–	–	–	–	–
1/4	–	0.250	6.350	0.2036	5.156	3	0.2130	5.410	–	28	–	–	–	–
5/16	–	0.312	7.938	0.2403	6.096	F	0.2570	6.527	18	–	–	–	–	–
5/16	–	0.312	7.938	0.2584	6.553	1	0.2720	6.908	–	24	–	–	–	–
–	M8	0.315	8.001	–	–	–	–	6.8	–	–	1.25	1.0	20	25
3/8	–	0.375	9.525	0.2938	7.442	5/16	0.3125	7.937	16	–	–	–	–	–
3/8	–	0.375	9.525	0.3209	8.153	Q	0.3320	8.432	–	24	–	–	–	–
–	M10	0.393	9.982	–	–	–	–	8.5	–	–	1.5	1.25	17	20
7/16	–	0.437	11.113	0.3447	8.738	U	0.3680	9.347	14	–	–	–	–	–
7/16	–	0.437	11.113	0.3726	9.448	25/64	0.3906	9.921	–	20	–	–	–	–
–	M12	0.471	11.963	–	–	–	–	10.30	–	–	1.75	1.25	14.5	20
1/2	–	0.500	12.700	0.4001	10.162	27/64	0.4219	10.715	13	–	–	–	–	–
1/2	–	0.500	12.700	0.4351	11.049	29/64	0.4531	11.509	–	20	–	–	–	–
–	M14	0.551	13.995	–	–	–	–	12.00	–	–	2	1.5	12.5	17
9/16	–	0.562	14.288	0.4542	11.531	31/64	0.4844	12.3031	12	–	–	–	–	–
9/16	–	0.562	14.288	0.4903	12.446	33/64	0.5156	13.096	–	18	–	–	–	–
5/8	–	0.625	15.875	0.5069	12.852	17/32	0.5312	13.493	11	–	–	–	–	–
5/8	–	0.625	15.875	0.5528	14.020	37/64	0.5781	14.684	–	18	–	–	–	–
–	M16	0.630	16.002	–	–	–	–	14.00	–	–	2	1.5	12.5	17
–	M18	0.709	18.008	–	–	–	–	15.50	–	–	2.5	1.5	10	17
3/4	–	0.750	19.050	0.6201	15.748	21/32	0.6562	16.668	10	–	–	–	–	–
3/4	–	0.750	19.050	0.6688	16.967	11/16	0.6875	17.462	–	16	–	–	–	–
–	M20	0.787	19.990	–	–	–	–	17.50	–	–	2.5	1.5	10	17
–	M22	0.866	21.996	–	–	–	–	19.50	–	–	2.5	1.5	10	17
7/8	–	0.875	22.225	0.7307	18.542	49/64	0.7656	19.446	9	–	–	–	–	–
7/8	–	0.875	22.225	0.7822	19.863	13/16	0.8125	20.637	–	14	–	–	–	–
–	M24	0.945	24.003	–	–	–	–	21.00	–	–	3	2	8.5	12.5
1	–	1.000	25.400	0.8376	21.2598	7/8	0.8750	22.225	8	–	–	–	–	–
1	–	1.000	25.400	0.8917	22.632	59/64	0.9219	23.415	–	12	–	–	–	–
–	M27	1.063	27.000	–	–	–	–	24.00	–	–	3	2	8.5	12.5

Source: Reprinted courtesy of The American Society of Mechanical Enginners

Table 3. ANSI Preferred Hole Basis Metric Clearance Fits

American National Standard Preferred Hole Basis Metric Clearance Fits (ANSI B4.2-1978, R1984)

Basic Size*		Loose Running			Free Running			Close Running			Sliding			Locational Clearance		
		Hole H11	Shaft c11	Fit†	Hole H9	Shaft d9	Fit†	Hole H8	Shaft f7	Fit†	Hole H7	Shaft g6	Fit†	Hole H7	Shaft h6	Fit†
1	Max	1.060	0.940	0.180	1.025	0.980	0.070	1.014	0.994	0.030	1.010	0.998	0.018	1.010	1.000	0.016
	Min	1.000	0.880	0.060	1.000	0.955	0.020	1.000	0.984	0.006	1.000	0.992	0.002	1.000	0.994	0.000
1.2	Max	1.260	1.140	0.180	1.225	1.180	0.070	1.214	1.194	0.030	1.210	1.198	0.018	1.210	1.200	0.016
	Min	1.200	1.080	0.060	1.200	1.155	0.020	1.200	1.184	0.006	1.200	1.192	0.002	1.200	1.194	0.000
1.6	Max	1.660	1.540	0.180	1.625	1.580	0.070	1.614	1.594	0.030	1.610	1.598	0.018	1.610	1.600	0.016
	Min	1.600	1.480	0.060	1.600	1.555	0.020	1.600	1.584	0.006	1.600	1.592	0.002	1.600	1.594	0.000
2	Max	2.060	1.940	0.180	2.025	1.980	0.070	2.014	1.994	0.030	2.010	1.998	0.018	2.010	2.000	0.016
	Min	2.000	1.880	0.060	2.000	1.955	0.020	2.000	1.984	0.006	2.000	1.992	0.002	2.000	1.994	0.000
2.5	Max	2.560	2.440	0.180	2.525	2.480	0.070	2.514	2.494	0.030	2.510	2.498	0.018	2.510	2.500	0.016
	Min	2.500	2.380	0.060	2.500	2.455	0.020	2.500	2.484	0.006	2.500	2.492	0.002	2.500	2.494	0.000
3	Max	3.060	2.940	0.180	3.025	2.980	0.070	3.014	2.994	0.030	3.010	2.998	0.018	3.010	3.000	0.016
	Min	3.000	2.880	0.060	3.000	2.955	0.020	3.000	2.984	0.006	3.000	2.992	0.002	3.000	2.994	0.000
4	Max	4.075	3.930	0.220	4.030	3.970	0.090	4.018	3.990	0.040	4.012	3.996	0.024	4.012	4.000	0.020
	Min	4.000	3.855	0.070	4.000	3.940	0.030	4.000	3.978	0.010	4.000	3.988	0.004	4.000	3.992	0.000
5	Max	5.075	4.930	0.220	5.030	4.970	0.090	5.018	4.990	0.040	5.012	4.996	0.024	5.012	5.000	0.020
	Min	5.000	4.855	0.070	5.000	4.940	0.030	5.000	4.978	0.010	5.000	4.988	0.004	5.000	4.992	0.000
6	Max	6.075	5.930	0.220	6.030	5.970	0.090	6.018	5.990	0.040	6.012	5.996	0.024	6.012	6.000	0.020
	Min	6.000	5.855	0.070	6.000	5.940	0.030	6.000	5.978	0.010	6.000	5.988	0.004	6.000	5.992	0.000
8	Max	8.090	7.920	0.260	8.036	7.960	0.112	8.022	7.987	0.050	8.015	7.995	0.029	8.015	8.000	0.024
	Min	8.000	7.830	0.080	8.000	7.924	0.040	8.000	7.972	0.013	8.000	7.986	0.005	8.000	7.991	0.000
10	Max	10.090	9.920	0.260	10.036	9.960	0.112	10.022	9.987	0.050	10.015	9.995	0.029	10.015	10.000	0.024
	Min	10.000	9.830	0.080	10.000	9.924	0.040	10.000	9.972	0.013	10.000	9.986	0.005	10.000	9.991	0.000
12	Max	12.110	11.905	0.315	12.043	11.956	0.136	12.027	11.984	0.061	12.018	11.994	0.035	12.018	12.000	0.029
	Min	12.000	11.795	0.095	12.000	11.907	0.050	12.000	11.966	0.016	12.000	11.983	0.006	12.000	11.989	0.000
16	Max	16.110	15.905	0.315	16.043	15.950	0.136	16.027	15.984	0.061	16.018	15.994	0.035	16.018	16.000	0.029
	Min	16.000	15.795	0.095	16.000	15.907	0.050	16.000	15.966	0.016	16.000	15.983	0.006	16.000	15.989	0.000
20	Max	20.130	19.890	0.370	20.052	19.935	0.169	20.033	19.980	0.074	20.021	19.993	0.041	20.021	20.000	0.034
	Min	20.000	19.760	0.110	20.000	19.883	0.065	20.000	19.959	0.020	20.000	19.980	0.007	20.000	19.987	0.000
25	Max	25.130	24.890	0.370	25.052	24.935	0.169	25.033	24.980	0.074	25.021	24.993	0.041	25.021	25.000	0.034
	Min	25.000	24.760	0.110	25.000	24.883	0.065	25.000	24.959	0.020	25.000	24.980	0.007	25.000	24.987	0.000
30	Max	30.130	29.890	0.370	30.052	29.935	0.169	30.033	29.980	0.074	30.021	29.993	0.041	30.021	30.000	0.034
	Min	30.000	29.760	0.110	30.000	29.883	0.065	30.000	29.959	0.020	30.000	29.980	0.007	30.000	29.987	0.000
40	Max	40.160	39.880	0.440	40.062	39.920	0.204	40.039	39.975	0.089	40.025	39.991	0.050	40.025	40.000	0.041
	Min	40.000	39.720	0.120	40.000	39.858	0.080	40.000	39.950	0.025	40.000	39.975	0.009	40.000	39.984	0.000
50	Max	50.160	49.870	0.450	50.062	49.920	0.204	50.039	49.975	0.089	50.025	49.991	0.050	50.025	50.000	0.041
	Min	50.000	49.710	0.130	50.000	49.858	0.080	50.000	49.950	0.025	50.000	49.975	0.009	50.000	49.984	0.000
60	Max	60.190	59.860	0.520	60.074	59.900	0.248	60.046	59.970	0.106	60.030	59.990	0.059	60.030	60.000	0.049
	Min	60.000	59.670	0.140	60.000	59.826	0.100	60.000	59.940	0.030	60.000	59.971	0.010	60.000	59.981	0.000
80	Max	80.190	79.850	0.530	80.074	79.900	0.248	80.046	79.970	0.106	80.030	79.990	0.059	80.030	80.000	0.049
	Min	80.000	79.660	0.150	80.000	79.826	0.100	80.000	79.940	0.030	80.000	79.971	0.010	80.000	79.981	0.000
100	Max	100.220	99.830	0.610	100.087	99.880	0.294	100.054	99.964	0.125	100.035	99.988	0.069	100.035	100.000	0.057
	Min	100.000	99.610	0.170	100.000	99.793	0.120	100.000	99.929	0.036	100.000	99.966	0.012	100.000	99.978	0.000
120	Max	120.220	119.820	0.620	120.087	119.880	0.294	120.054	119.964	0.125	120.035	119.988	0.069	120.035	120.000	0.057
	Min	120.000	119.600	0.180	120.000	119.793	0.120	120.000	119.929	0.036	120.000	119.966	0.012	120.000	119.978	0.000
160	Max	160.250	159.790	0.710	160.100	159.855	0.345	160.063	159.957	0.146	160.040	159.986	0.079	160.040	160.000	0.065
	Min	160.000	159.540	0.210	160.000	159.755	0.145	160.000	159.917	0.043	160.000	159.961	0.014	160.000	159.975	0.000
200	Max	200.290	199.760	0.820	200.115	199.830	0.400	200.072	199.950	0.168	200.046	199.985	0.090	200.046	200.000	0.075
	Min	200.000	199.470	0.240	200.000	199.715	0.170	200.000	199.904	0.050	200.000	199.956	0.015	200.000	199.971	0.000
250	Max	250.290	249.720	0.860	250.115	249.830	0.400	250.072	249.950	0.168	250.046	249.985	0.090	250.046	250.000	0.075
	Min	250.000	249.430	0.280	250.000	249.715	0.170	250.000	249.904	0.050	250.000	249.956	0.015	250.000	249.971	0.000
300	Max	300.320	299.670	0.970	300.130	299.810	0.450	300.081	299.944	0.189	300.052	299.983	0.101	300.052	300.000	0.084
	Min	300.000	299.350	0.330	300.000	299.680	0.190	300.000	299.892	0.056	300.000	299.951	0.017	300.000	299.968	0.000
400	Max	400.360	399.600	1.120	400.140	399.790	0.490	400.089	399.938	0.208	400.057	399.982	0.011	400.057	400.000	0.093
	Min	400.000	399.240	0.400	400.000	399.650	0.210	400.000	399.881	0.062	400.000	399.946	0.018	400.000	399.964	0.000
500	Max	500.400	499.520	1.280	500.155	499.770	0.540	500.097	499.932	0.228	500.063	499.980	0.123	500.063	500.000	
	Min	500.000	499.120	0.480	500.000	499.615	0.230	500.000	499.869	0.068	500.000	499.940	0.020	500.000		

All dimensions are in millimeters.
*The sizes shown are first choice basic sizes (see Table 1). Preferred fits for other sizes can be calculated from data given in ANSI B4
†All fits shown in this table have clearance.
Source: Reprinted courtesy of The American Society of Mechanical Engineers

Table 4. ANSI Preferred Hole Basis Transition and Interference Fits

American National Standard Preferred Hole Basis Metric Transition and Interference Fits (ANSI B4.2-1978, RI984)

Basic Size*		Locational Transition			Locational Transition			Locational Interference			Medium Drive			Force		
		Hole H7	Shaft k6	Fit†	Hole H7	Shaft n6	Fit†	Hole H7	Shaft p6	Fit†	Hole H7	Shaft s6	Fit†	Hole H7	Shaft u6	Fit†
1	Max	1.010	1.006	+0.010	1.010	1.010	+0.006	1.010	1.012	+0.004	1.010	1.020	−0.004	1.010	1.024	−0.008
	Min	1.000	1.000	−0.006	1.000	1.004	−0.010	1.000	1.006	−0.012	1.000	1.014	−0.020	1.000	1.018	−0.024
1.2	Max	1.210	1.206	+0.010	1.210	1.210	+0.006	1.210	1.212	+0.004	1.210	1.220	−0.004	1.210	1.224	−0.008
	Min	1.200	1.200	−0.006	1.200	1.204	−0.010	1.200	1.206	−0.012	1.200	1.214	−0.020	1.200	1.218	−0.024
1.6	Max	1.610	1.606	+0.010	1.610	1.610	+0.006	1.610	1.612	+0.004	1.610	1.620	−0.004	1.610	1.624	−0.008
	Min	1.600	1.600	−0.006	1.600	1.604	−0.010	1.600	1.606	−0.012	1.600	1.614	−0.020	1.600	1.618	−0.024
2	Max	2.010	2.006	+0.010	2.010	2.010	+0.006	2.010	2.012	+0.004	2.010	2.020	−0.004	2.010	2.024	−0.008
	Min	2.000	2.000	−0.006	2.000	2.004	−0.010	2.000	2.006	−0.012	2.000	2.014	−0.020	2.000	2.018	−0.024
2.5	Max	2.510	2.506	+0.010	2.510	2.510	+0.006	2.510	2.512	+0.004	2.510	2.520	−0.004	2.510	2.524	−0.008
	Min	2.500	2.500	−0.006	2.500	2.504	−0.010	2.500	2.506	−0.012	2.500	2.514	−0.020	2.500	2.518	−0.024
3	Max	3.010	3.006	+0.010	3.010	3.010	+0.006	3.010	3.012	+0.004	3.010	3.020	−0.004	3.010	3.024	−0.008
	Min	3.000	3.000	−0.006	3.000	3.004	−0.010	3.000	3.006	−0.012	3.000	3.014	−0.020	3.000	3.018	−0.024
4	Max	4.012	4009	+0.011	4.012	4.016	+0.004	4.012	4.020	0.000	4.012	4.027	−0.007	4.012	4.031	−0.011
	Min	4.000	4.001	−0.009	4.000	4.008	−0.016	4.000	4.012	−0.020	4.000	4.019	−0.027	4.000	4.023	−0.031
5	Max	5.012	5.009	+0.011	5.012	5.016	+0.004	5.012	5.020	0.000	5.012	5.027	−0.007	5.012	5.031	−0.011
	Min	5.000	5.001	−0.009	5.000	5.008	−0.016	5.000	5.012	−0.020	5.000	5.019	−0.027	5.000	5.023	−0.031
6	Max	6.012	6.009	+0.011	6.012	6.016	+0.004	6.012	6.020	0.000	6.012	6.027	−0.007	6.012	6.031	−0.011
	Min	6.000	6.001	−0.009	6.000	6.008	−0.016	6.000	6.012	−0.020	6.000	6.019	−0.027	6.000	6.023	−0.031
8	Max	8.015	8.010	+0.014	8.015	8.019	+0.005	8.015	8.024	0.000	8.015	8.032	−0.008	8.015	8.037	−0.013
	Min	8.000	8.001	−0.010	8.000	8.010	−0.019	8.000	8.015	−0.024	8.000	8.023	−0.032	8.000	8.028	−0.037
10	Max	10.015	10.010	+0.014	10.015	10.019	+0.005	10.015	10.024	0.000	10.015	10.032	−0.008	10.015	10.037	−0.013
	Min	10.000	10.001	−0.010	10.000	10.010	−0.019	10.000	10.015	−0.024	10.000	10.023	−0.032	10.000	10.028	−0.037
12	Max	12.018	12.012	+0.017	12.018	12.023	+0.006	12.018	12.029	0.000	12.018	12.039	−0.010	12.018	12.044	−0.015
	Min	12.000	12.001	−0.012	12.000	12.012	−0.023	12.000	12.018	−0.029	12.000	12.028	−0.039	12.000	12.033	−0.044
16	Max	16.018	16.012	+0.017	16.018	16.023	+0.006	16.018	16.029	0.000	16.018	16.039	−0.010	16.018	16.044	−0.015
	Min	16.000	16.001	−0.012	16.000	16.012	−0.023	16.000	16.018	−0.029	16.000	16.028	−0.039	16.000	16.033	−0.044
20	Max	20.021	20.015	+0.019	20.021	20.028	+0.006	20.021	20.035	−0.001	20.021	20.048	−0.014	20.021	20.054	−0.020
	Min	20.000	20.002	−0.015	20.000	20.015	−0.028	20.000	20.022	−0.035	20.000	20.035	−0.048	20.000	20.041	−0.054
25	Max	25.021	25.015	+0.019	25.021	25.028	+0.006	25.021	25.035	−0.001	25.021	25.048	−0.014	25.021	25.061	−0.027
	Min	25.000	25.002	−0.015	25.000	25.015	−0.028	25.000	25.022	−0.035	25.000	25.035	−0.048	25.000	25.048	−0.061
30	Max	30.021	30.015	+0.019	30.021	30.028	+0.006	30.021	30.035	−0.001	30.021	30.048	−0.014	30.021	30.061	−0.027
	Min	30.000	30.002	−0.015	30.000	30.015	−0.028	30.000	30.022	−0.035	30.000	30.035	−0.048	30.000	30.048	−0.061
40	Max	40.025	40.018	+0.023	40.025	40.033	+0.008	40.025	40.042	−0.001	40.025	40.059	−0.018	40.025	40.076	−0.035
	Min	40.000	40.002	−0.018	40.000	40.017	−0.033	40.000	40.026	−0.042	40.000	40.043	−0.059	40.000	40.060	−0.076
50	Max	50.025	50.018	+0.023	50.025	50.033	+0.008	50.025	50.042	−0.001	50.025	50.059	−0.018	50.025	50.086	−0.045
	Min	50.000	50.002	−0.018	50.000	50.017	−0.033	50.000	50.026	−0.042	50.000	50.043	−0.059	50.000	50.070	−0.086
60	Max	60.030	60.021	+0.028	60.030	60.039	+0.010	60.030	60.051	−0.002	60.030	60.072	−0.023	60.030	60.106	−0.057
	Min	60.000	60.002	−0.021	60.000	60.020	−0.039	60.000	60.032	−0.051	60.000	60.053	−0.072	60.000	60.087	−0.106
80	Max	80.030	80.021	+0.028	80.030	80.039	+0.010	80.030	80.051	−0.002	80.030	80.078	−0.029	80.030	80.121	−0.072
	Min	80.000	80.002	−0.021	80.000	80.020	−0.039	80.000	80.032	−0.051	80.000	80.059	−0.078	80.000	80.102	−0.121
100	Max	100.035	100.025	+0.032	100.035	100.045	+0.012	100.035	100.059	−0.002	100.035	100.093	−0.036	100.035	100.146	−0.089
	Min	100.000	100.003	−0.025	100.000	100.023	−0.045	100.000	100.037	−0.059	100.000	100.071	−0.093	100.000	100.124	−0.146
120	Max	120.035	120.025	+0.032	120.035	120.045	+0.012	120.035	120.059	−0.002	120.035	120.101	−0.044	120.035	120.166	−0.109
	Min	120.000	120.003	−0.025	120.000	120.023	−0.045	120.000	120.037	−0.059	120.000	120.079	−0.101	120.000	120.144	−0.166
160	Max	160.040	160.028	+0.037	160.040	160.052	+0.013	160.040	160.068	−0.003	160.040	160.125	−0.060	160.040	160.215	−0.150
	Min	160.000	160.003	−0.028	160.000	160.027	−0.052	160.000	160.043	−0.068	160.000	160.100	−0.125	160.000	160.190	−0.215
200	Max	200.046	200.033	+0.042	200.046	200.060	+0.015	200.046	200.079	−0.004	200.046	200.151	−0.076	200.046	200.265	−0.190
	Min	200.000	200.004	−0.033	200.000	200.031	−0.060	200.000	200.050	−0.079	200.000	200.122	−0.151	200.000	200.236	−0.265
250	Max	250.046	250.033	+0.042	250.046	250.060	+0.015	250.046	250.079	−0.004	250.046	250.169	−0.094	250.046	250.313	−0.238
	Min	250.000	250.004	−0.033	250.000	250.031	−0.060	250.000	250.050	−0.079	250.000	250.140	−0.169	250.000	250.284	−0.313
300	Max	300.052	300.036	+0.048	300.052	300.066	+0.018	300.052	300.088	−0.004	300.052	300.202	−0.118	300.052	300.382	−0.298
	Min	300.000	300.004	−0.036	300.000	300.034	−0.066	300.000	300.056	−0.088	300.000	300.170	−0.202	300.000	300.350	−0.382
400	Max	400.057	400.040	+0.053	400.057	400.073	+0.020	400.057	400.098	−0.005	400.057	400.244	−0.151	400.057	400.471	−0.378
	Min	400.000	400.004	−0.040	400.000	400.037	−0.073	400.000	400.062	−0.098	400.000	400.208	−0.244	400.000	400.435	−0.471
500	Max	500.063	500.045	+0.058	500.063	500.080	+0.023	500.063	500.108	−0.005	500.063	500.292	−0.189	500.063	500.580	−0.477
	Min	500.000	500.005	−0.045	500.000	500.040	−0.080	500.000	500.068	−0.108	500.000	500.252	−0.292	500.000	500.540	−0.580

*...re in millimeters.
...e first choice basic sizes (see Table 1). Preferred fits for other sizes can be calculated from data given in ANSI B4.2-1978 (R1984).
...earance; a minus sign indicates interference.
...of The American Society of Mechanical Engineers.

Table 5. ANSI Preferred Shaft Basis Metric Clearance Fits

American National Standard Preferred Shaft Basis Metric Clearance Fits (ANSI B4.2-1978, R1984)

Basic Size*		Loose Running Hole C11	Loose Running Shaft h11	Fit†	Free Running Hole D9	Free Running Shaft h9	Fit†	Close Running Hole F8	Close Running Shaft h7	Fit†	Sliding Hole G7	Sliding Shaft h6	Fit†	Locational Clearance Hole H7	Locational Clearance Shaft h6	Fit†
1	Max	1.120	1.000	0.180	1.045	1.000	0.070	1.020	1.000	0.030	1.012	1.000	0.018	1.010	1.000	0.016
	Min	1.060	0.940	0.060	1.020	0.975	0.020	1.006	0.990	0.006	1.002	0.994	0.002	1.000	0.994	0.000
1.2	Max	1.320	1.200	0.180	1.245	1.200	0.070	1.220	1.200	0.030	1.212	1.200	0.018	1.210	1.200	0.016
	Min	1.260	1.140	0.060	1.220	1.175	0.020	1.206	1.190	0.006	1.202	1.194	0.002	1.200	1.194	0.000
1.6	Max	1.720	1.600	0.180	1.645	1.600	0.070	1.620	1.600	0.030	1.612	1.600	0.018	1.610	1.600	0.016
	Min	2.660	1.540	0.060	1.620	1.575	0.020	1.606	1.590	0.006	1.602	1.594	0.002	1.600	1.594	0.000
2	Max	2.120	2.000	0.180	2.045	2.000	0.070	2.020	2.000	0.030	2.012	2.000	0.018	2.010	2.000	0.016
	Min	2.060	1.940	0.060	2.020	1.975	0.020	2.006	1.990	0.006	2.002	1.994	0.002	2.000	1.994	0.000
2.5	Max	2.620	2.500	0.180	2.545	2.500	0.070	2.520	2.500	0.030	2.512	2.500	0.018	2.510	2.500	0.016
	Min	2.560	2.440	0.060	2.520	2.475	0.020	2.506	2.490	0.006	2.502	2.494	0.001	2.500	2.494	0.000
3	Max	3.120	3.000	0.180	3.045	3.000	0.070	3.020	3.000	0.030	3.012	3.000	0.018	3.010	3.000	0.016
	Min	3.060	2.940	0.060	3.020	2.975	0.020	3.006	2.990	0.006	3.002	2.994	0.002	3.000	2.994	0.000
4	Max	4.145	4.000	0.220	4.060	4.000	0.090	4.028	4.000	0.040	4.016	4.000	0.024	4.012	4.000	0.020
	Min	4.070	3.925	0.070	4.030	3.970	0.030	4.010	3.988	0.010	4.004	3.992	0.004	4.000	3.992	0.000
5	Max	5.145	5.000	0.220	5.060	5.000	0.090	5.028	5.000	0.040	5.016	5.000	0.024	5.012	5.000	0.020
	Min	5.070	4.925	0.070	5.030	4.970	0.030	5.010	4.988	0.010	5.004	4.992	0.004	5.000	4.992	0.000
6	Max	6.145	6.000	0.220	6.060	6.000	0.090	6.028	6.000	0.040	6.016	6.000	0.024	6.012	6.000	0.020
	Min	6.070	5.925	0.070	6.030	5.970	0.030	6.010	5.988	0.010	6.004	5.992	0.004	6.000	5.992	0.000
8	Max	8.170	8.000	0.260	8.076	8.000	0.112	8.035	8.000	0.050	8.020	8.000	0.029	8.015	8.000	0.024
	Min	8.080	7.910	0.080	8.040	7.964	0.040	8.013	7.985	0.013	8.005	7.991	0.005	8.000	7.991	0.000
10	Max	10.170	10.000	0.260	10.076	10.000	0.112	10.035	10.000	0.050	10.020	10.000	0.029	10.015	10.000	0.024
	Min	10.080	9.910	0.080	10.040	9.964	0.040	10.013	9.985	0.013	10.005	9.991	0.005	10.000	9.991	0.000
12	Max	12.205	12.000	0.315	12.093	12.000	0.136	12.043	12.000	0.061	12.024	12.000	0.035	12.018	12.000	0.029
	Min	12.095	11.890	0.095	12.050	11.957	0.050	12.016	11.982	0.016	12.006	11.989	0.006	12.000	11.989	0.000
16	Max	16.205	16.000	0.315	16.093	16.000	0.136	16.043	16.000	0.061	16.024	16.000	0.035	16.018	16.000	0.029
	Min	16.095	15.890	0.095	16.050	15.957	0.050	16.016	15.982	0.016	16.006	15.989	0.006	16.000	15.989	0.000
20	Max	20.240	20.000	0.370	20.117	20.000	0.169	20.053	20.000	0.074	20.028	20.000	0.041	20.021	20.000	0.034
	Min	20.110	19.870	0.110	20.065	19.948	0.065	20.020	19.979	0.020	20.007	19.987	0.007	20.000	19.987	0.000
25	Max	25.240	25.000	0.370	25.117	25.000	0.169	25.053	25.000	0.074	25.028	25.000	0.041	25.021	25.000	0.034
	Min	25.110	24.870	0.110	25.065	24.948	0.065	25.020	24.979	0.020	25.007	24.987	0.007	25.000	24.987	0.000
30	Max	30.240	30.000	0.370	30.117	30.000	0.169	30.053	30.000	0.074	30.028	30.000	0.041	30.021	30.000	0.034
	Min	30.110	29.870	0.110	30.065	29.948	0.065	30.020	29.979	0.020	30.007	29.987	0.007	30.000	29.987	0.000
40	Max	40.280	40.000	0.440	40.142	40.000	0.204	40.064	40.000	0.089	40.034	40.000	0.050	40.025	40.000	0.041
	Min	40.120	39.840	0.120	40.080	39.938	0.080	40.025	39.975	0.025	40.009	39.984	0.009	40.000	39.984	0.000
50	Max	50.290	50.000	0.450	50.142	50.000	0.204	50.064	50.000	0.089	50.034	50.000	0.050	50.025	50.000	0.041
	Min	50.130	49.840	0.130	50.080	49.938	0.080	50.025	49.975	0.025	50.009	49.984	0.009	50.000	49.984	0.000
60	Max	60.330	60.000	0.520	60.174	60.000	0.248	60.076	60.000	0.106	60.040	60.000	0.059	60.030	60.000	0.049
	Min	60.140	59.810	0.140	60.100	59.926	0.100	60.030	59.970	0.030	60.010	59.981	0.010	60.000	59.981	0.000
80	Max	80.340	80.000	0.530	80.174	80.000	0.248	80.076	80.000	0.106	80.040	80.000	0.059	80.030	80.000	0.049
	Min	80.150	79.810	0.150	80.100	79.926	0.100	80.030	79.970	0.030	80.010	79.981	0.010	80.000	79.981	0.000
100	Max	100.390	100.000	0.610	100.207	100.000	0.294	100.090	100.000	0.125	100.047	100.000	0.069	100.035	100.000	0.057
	Min	100.170	99.780	0.170	100.120	99.913	0.120	100.036	99.965	0.036	100.012	99.978	0.012	100.000	99.978	0.000
120	Max	120.400	120.000	0.620	120.207	120.000	0.294	120.090	120.000	0.125	120.047	120.000	0.069	120.035	120.000	0.057
	Min	120.180	119.780	0.180	120.120	119.913	0.120	120.036	119.965	0.036	120.012	119.978	0.012	120.000	119.978	0.000
160	Max	160.460	160.000	0.710	160.245	260.000	0.345	160.106	160.000	0.146	160.054	160.000	0.079	160.040	160.000	0.065
	Min	160.210	159.750	0.210	160.145	159.900	0.145	160.043	159.960	0.043	160.014	159.975	0.014	160.000	159.975	0.000
200	Max	200.530	200.000	0.820	200.285	200.000	0.400	200.122	200.000	0.168	200.061	200.000	0.090	200.046	200.000	0.075
	Min	200.240	199.710	0.240	200.170	199.885	0.170	200.050	199.954	0.050	200.015	199.971	0.015	200.000	199.971	0.000
250	Max	250.570	250.000	0.860	250.285	250.000	0.400	250.122	250.000	0.168	250.061	250.000	0.090	250.046	250.000	0.075
	Min	250.280	249.710	0.280	250.170	249.885	0.170	250.050	249.954	0.050	250.015	249.971	0.015	250.000	249.971	0.000
300	Max	300.650	300.000	0.970	300.320	300.000	0.450	300.137	300.000	0.189	300.069	300.000	0.101	300.052	300.000	0.084
	Min	300.330	299.680	0.330	300.190	299.870	0.190	300.056	299.948	0.056	300.017	299.968	0.017	300.000	299.968	0.000
400	Max	400.760	400.000	1.120	400.350	400.000	0.490	400.151	400.000	0.208	400.075	400.000	0.111	400.057	400.000	0.093
	Min	400.400	399.640	0.400	400.210	399.860	0.210	400.062	399.943	0.062	400.018	399.964	0.018	400.000	399.964	0.000
500	Max	500.880	500.000	1.280	500.385	500.000	0.540	500.165	500.000	0.228	500.083	500.000	0.123	500.063	500.000	0.103
	Min	500.480	499.600	0.480	500.230	499.845	0.230	500.068	499.937	0.068	500.020	499.960	0.020	500.000	499.960	0.000

All dimensions are in millimeters.
*The sizes shown are first choice basic sizes (see Table 1). Preferred fits for other sizes can be calculated from data given in ANSI B4.2-1978 (RI984).
†All fits shown in this table have clearance.
Source: Reprinted courtesy of The American Society of Mechanical Engineers.

Table 6. ANSI Preferred Shaft Basis Metric Transition and Interference Fits

American National Standard Preferred Shaft Basis Metric Transition and Interference Fits (ANSI B4.2-1978, R1984)

Basic Size*		Locational Transition			Locational Transition			Locational Interference			Medium Drive			Force		
		Hole K7	Shaft h6	Fit†	Hole N7	Shaft h6	Fit†	Hole P7	Shaft h6	Fit†	Hole S7	Shaft h6	Fit†	Hole U7	Shaft h6	Fit†
1	Max	1.000	1.000	+0.006	0.996	1.000	+0.002	0.994	1.000	0.000	0.986	1.000	−0.008	0.982	1.000	−0.012
	Min	0.990	0.994	−0.010	0.986	0.954	−0.014	0.984	0.994	−0.016	0.976	0.994	−0.024	0.972	0.994	−0.028
1.2	Max	1.200	1.200	+0.006	1.196	1.200	+0.002	1.194	1.200	0.000	1.186	1.200	−0.008	1.182	1.200	−0.012
	Min	1.190	1.194	−0.010	1.186	1.194	−0.014	1.184	1.194	−0.016	1.176	1.194	−0.024	1.172	1.194	−0.028
1.6	Max	1.600	1.600	+0.006	1.596	1.600	+0.002	1.594	1.600	0.000	1.586	1.600	−0.008	1.582	1.600	−0.012
	Min	1.590	1.594	−0.010	1.586	1.594	−0.014	1.584	1.594	−0.016	1.576	1.594	−0.024	1.572	1.594	−0.028
2	Max	2.000	2.000	+0.006	1.996	2.000	+0.002	1.994	2.000	0.000	1.986	2.000	−0.008	1.982	2.000	−0.012
	Min	1.990	1.994	−0.010	1.986	1.994	−0.014	1.984	1.994	−0.016	1.976	1.994	−0.024	1.972	1.994	−0.028
2.5	Max	2.500	2.500	+0.006	2.496	2.500	+0.002	2.494	2.500	0.000	2.486	2.500	−0.008	2.482	2.500	−0.012
	Min	2.490	2.494	−0.010	2.486	2.494	−0.014	2.484	2.494	−0.016	2.476	2.494	−0.024	2.472	2.494	−0.028
3	Max	3.000	3.000	+0.006	2.996	3.000	+0.002	2.994	3.000	0.000	2.986	3.000	−0.008	2.982	3.000	−0.012
	Min	2.990	2.994	−0.010	2.986	2.994	−0.014	2.984	2.994	−0.016	2.976	2.994	−0.024	2.972	2.994	−0.028
4	Max	4.003	4.000	+0.011	3.996	4.000	+0.004	3.992	4.000	0.000	3.985	4.000	−0.007	3.981	4.000	−0.011
	Min	3.991	3.992	−0.009	3.984	3.992	−0.016	3.980	3.992	−0.020	3.973	3.992	−0.027	3.969	3.992	−0.031
5	Max	5.003	5.000	+0.011	4.996	5.000	+0.004	4.992	5.000	0.000	4.985	5.000	−0.007	4.981	5.000	−0.011
	Min	4.991	4.992	−0.009	4.984	4.992	−0.016	4.980	4.992	−0.020	4.973	4.992	−0.027	4.969	4.992	−0.031
6	Max	6.003	6.000	+0.011	5.996	6.000	+0.004	5.992	6.000	0.000	5.985	6.000	−0.007	5.981	6.000	−0.011
	Min	5.991	5.992	−0.009	5.984	5.992	−0.016	5.980	5.992	−0.020	5.973	5.992	−0.027	5.969	5.992	−0.031
8	Max	8.005	8.000	+0.014	7.996	8.000	+0.005	7.991	8.000	0.000	7.983	8.000	−0.008	7.978	8.000	−0.013
	Min	7.990	7.991	−0.010	7.981	7.991	−0.019	7.976	7.991	−0.024	7.968	7.991	−0.032	7.963	7.991	−0.037
10	Max	10.005	10.000	+0.014	9.996	10.000	+0.005	9.991	10.000	0.000	9.983	10.000	−0.008	9.978	10.000	−0.013
	Min	9.990	9.991	−0.010	9.981	9.991	−0.019	9.976	9.991	−0.024	9.968	9.991	−0.032	9.963	9.991	−0.037
12	Max	12.006	12.000	+0.017	11.995	12.000	+0.006	11.989	12.000	0.000	11.979	12.000	−0.010	11.974	12.000	−0.015
	Min	11.988	11.989	−0.012	11.977	11.989	−0.023	11.971	11.989	−0.029	11.961	11.989	−0.039	11.956	11.989	−0.044
16	Max	16.006	16.000	+0.017	15.995	16.000	+0.006	15.989	16.000	0.000	15.979	16.000	−0.010	15.974	16.000	−0.015
	Min	15.988	15.989	−0.012	15.977	15.989	−0.023	15.971	15.989	−0.029	15.961	15.989	−0.039	15.956	15.989	−0.044
20	Max	20.006	20.000	+0.019	19.993	20.000	+0.006	19.986	20.000	−0.001	19.973	20.000	−0.014	19.967	20.000	−0.020
	Min	19.985	19.987	−0.015	19.972	19.987	−0.028	19.965	19.987	−0.035	19.952	19.987	−0.048	19.946	19.987	−s0.054
25	Max	25.006	25.000	+0.019	24.993	25.000	+0.006	24.986	25.000	−0.001	24.973	25.000	−0.014	24.960	25.000	−0.027
	Min	24.985	24.987	−0.015	24.972	24.987	−0.028	24.965	24.987	−0.035	24.952	24.987	−0.048	24.939	24.987	−0.061
30	Max	30.006	30.000	+0.019	29.993	30.000	+0.006	29.986	30.000	−0.001	29.973	30.000	−0.014	29.960	30.000	−0.027
	Min	29.985	29.987	−0.015	29.972	29.987	−0.028	29.965	29.987	−0.035	29.952	29.987	−0.048	29.939	29.987	−0.061
40	Max	40.007	40.000	+0.023	39.992	40.000	+0.008	39.983	40.000	−0.001	39.966	40.000	−0.018	39.949	40.000	−0.035
	Min	39.982	39.984	−0.018	39.967	39.984	−0.033	39.958	39.984	−0.042	39.941	39.984	−0.059	39.924	39.984	−0.076
50	Max	50.007	50.000	−0.023	49.992	50.000	+0.008	49.983	50.000	−0.001	49.966	50.000	−0.018	49.939	50.000	−0.045
	Min	49.982	49.984	−0.018	49.967	49.984	−0.033	49.958	49.984	−0.042	49.941	49.984	−0.059	49.914	49.984	−0.086
60	Max	60.009	60.000	+0.028	59.991	60.000	+0.010	59.979	60.000	−0.002	59.958	60.000	−0.023	59.924	60.000	−0.087
	Min	59.979	59.981	−0.021	59.961	59.981	−0.039	59.949	59.981	−0.051	59.928	59.981	−0.072	59.894	59.981	−0.106
80	Max	80.009	80.000	+0.028	79.991	80.000	+0.010	79.979	80.000	−0.002	79.952	80.000	−0.029	79.909	80.000	−0.072
	Min	79.979	79.981	−0.021	79.961	79.981	−0.039	79.949	79.981	−0.051	79.922	79.981	−0.078	79.879	79.981	−0.121
100	Max	100.010	100.000	+0.032	99.990	100.000	+0.012	99.976	100.000	−0.002	99.942	100.000	−0.036	99.889	100.000	−0.089
	Min	99.975	99.978	−0.025	99.955	99.978	−0.045	99.941	99.978	−0.059	99.907	99.978	−0.093	99.854	99.978	−0.146
120	Max	120.010	120.000	+0.032	119.990	120.000	+0.012	119.976	120.000	−0.002	119.934	120.000	−0.044	119.869	120.000	−0.109
	Min	119.975	119.978	−0.025	119.955	119.978	−0.045	119.941	119.978	−0.059	119.899	119.978	−0.101	119.834	119.978	−0.166
160	Max	160.012	160.000	+0.037	159.988	160.000	+0.013	159.972	160.000	−0.003	159.915	160.000	−0.060	159.825	160.000	−0.150
	Min	159.972	159.975	−0.028	159.948	159.975	−0.052	159.932	159.975	−0.068	159.875	159.975	−0.125	159.785	159.975	−0.215
200	Max	200.013	200.000	+0.042	199.986	200.000	+0.015	199.967	200.000	−0.004	199.895	200.000	−0.076	199.781	200.000	−0.190
	Min	199.967	199.971	−0.033	199.940	199.971	−0.060	199.921	199.971	−0.079	199.849	199.971	−0.151	199.735	199.971	−0.265
250	Max	250.013	250.000	+0.042	249.986	250.000	+0.015	249.967	250.000	−0.004	249.877	250.000	−0.094	249.733	250.000	−0.238
	Min	249.967	249.971	−0.033	249.940	249.971	−0.060	249.921	249.971	−0.079	249.831	249.971	−0.169	249.687	249.971	−0.313
300	Max	300.016	300.000	+0.048	299.986	300.000	+0.018	299.964	300.000	−0.004	299.850	300.000	−0.118	299.670	300.000	−0.298
	Min	299.964	299.968	−0.036	299.934	299.968	−0.066	299.912	299.968	−0.088	299.798	299.968	−0.202	299.618	299.968	−0.382
400	Max	400.017	400.000	+0.053	399.984	400.000	+0.020	399.959	400.000	−0.005	399.813	400.000	−0.151	399.586	400.000	−0.378
	Min	399.960	399.964	−0.040	399.927	399.964	−0.073	399.902	399.964	−0.098	399.756	399.964	−0.244	399.529	399.964	−0.471
500	Max	500.018	500.000	+0.058	499.983	500.000	+0.023	499.955	500.000	−0.005	499.771	500.000	−0.189	499.483	500.000	−0.477
	Min	499.955	499.960	−0.045	499.920	499.960	−0.080	499.892	499.960	−0.108	499.708	499.960	−0.292	499.420	499.960	−0.580

All dimensions are in millimeters.
*The sizes shown are first choice basic sizes (see Table 1). Preferred fits for other sizes can be calculated from data given in ANSI B4.2-1978 (R1984).
†A plus sign indicates clearance: a minus sign indicates interference.
Source: Reprinted courtesy of The American Society of Mechanical Engineers.

Table 7. Description of Preferred Metric Fits

	ISO Symbol		Description
	Hole Basis	Shaft Basis	
Clearance Fits ↑ More Clearance	H11/c11	C11/h11	**Loose running** fit for wide commercial tolerances or allowances on external members.
	H9/d9	D9/h9	**Free running** fit not for use where accuracy is essential, but good for large temperature variations, high running speeds, or heavy journal pressures.
	H8/f7	F8/h7	**Close running** fit for running on accurate machines and for accurate location at moderate speeds and journal pressures.
	H7/g6	G7/h6	**Sliding fit** not intended to run freely, but to move and turn freely and locate accurately.
	H7/h6	H7/h6	**Locational clearance** fit provides snug fit for locating stationary parts; but can be freely assembled and disassembled.
Transition Fits	H7/k6	K7/h6	**Locational transition** fit for accurate location, a compromise between clearance and interference.
	H7/n6	N7/h6	**Locational transition** fit for more accurate location where greater interference is permissible.
Interference Fits ↓ More Interference	H7/p6[1]	P7/h6	**Locational interference** fit for parts requiring rigidity and alignment with prime accuracy of location but without special bore pressure requirements.
	H7/s6	S7/h6	**Medium drive** fit for ordinary steel parts or shrink fits on light sections, the tightest fit usable with cast iron.
	H7/u6	U7/h6	**Force** fit suitable for parts which can be highly stressed or for shrink fits where the heavy pressing forces required are impractical.

[1] Transition fits for basic sizes in range from 0 through 3 mm.
Source: Reprinted courtesy of The American Society of Mechanical Engineers.

Table 8. ANSI Force and Shrink Fits (FN)

ANSI Standard Force and Shrink Fits (ANSI B4.1-1967, R1979)

Nominal Size Range, Inches Over To	Class FN 1 Inter-ference*	Standard Tolerance Limits Hole H6	Shaft	Class FN 2 Inter-ference*	Standard Tolerance Limits Hole H7	Shaft s6	Class FN 3 Inter-ference*	Standard Tolerance Limits Hole H7	Shaft t6	Class FN 4 Inter-ference*	Standard Tolerance Limits Hole H7	Shaft u6	Class FN 5 Inter-ference*	Standard Tolerance Limits Hole H8	Shaft x7
					Values shown below are in thousandths of an inch										
0–0.12	0.05 / 0.5	+0.25 / 0	+0.5 / +0.3	0.2 / 0.85	+0.4 / 0	+0.85 / +0.6				0.3 / 0.95	+0.4 / 0	+0.95 / +0.7	0.3 / 1.3	+0.6 / 0	+1.3 / +0.9
0.12–0.24	0.1 / 0.6	+0.3 / 0	+0.6 / +0.4	0.2 / 1.0	+0.5 / 0	+1.0 / +0.7				0.4 / 1.2	+0.5 / 0	+1.2 / +0.9	0.5 / 1.7	+0.7 / 0	+1.7 / +1.2
0.24–0.40	0.1 / 0.75	+0.4 / 0	+0.75 / +0.5	0.4 / 1.4	+0.6 / 0	+1.4 / +1.0				0.6 / 1.6	+0.6 / 0	+1.6 / +1.2	0.5 / 2.0	+0.9 / 0	+2.0 / +1.4
0.40–0.56	0.1 / 0.8	+0.4 / 0	+0.8 / +0.5	0.5 / 1.6	+0.7 / 0	+1.6 / +1.2				0.7 / 1.8	+0.7 / 0	+1.8 / +1.4	0.6 / 2.3	+1.0 / 0	+2.3 / +1.6
0.56–0.71	0.2 / 0.9	+0.4 / 0	+0.9 / +0.6	0.5 / 1.6	+0.7 / 0	+1.6 / +1.2				0.7 / 1.8	+0.7 / 0	+1.8 / +1.4	0.8 / 2.5	+1.0 / 0	+2.5 / +1.8
0.71–0.95	0.2 / 1.1	+0.5 / 0	+1.1 / +0.7	0.6 / 1.9	+0.8 / 0	+1.9 / +1.4				0.8 / 2.1	+0.8 / 0	+2.1 / +1.6	1.0 / 3.0	+1.2 / 0	+3.0 / +2.2
0.95–1.19	0.3 / 1.2	+0.5 / 0	+1.2 / +0.8	0.6 / 1.9	+0.8 / 0	+1.9 / +1.4	0.8 / 2.1	+0.8 / 0	+2.1 / +1.6	1.0 / 2.3	+0.8 / 0	+2.3 / +1.8	1.3 / 3.3	+1.2 / 0	+3.3 / +2.5
1.19–1.58	0.3 / 1.3	+0.6 / 0	+1.3 / +0.9	0.8 / 2.4	+1.0 / 0	+2.4 / +1.8	1.0 / 2.6	+1.0 / 0	+2.6 / +2.0	1.5 / 3.1	+1.0 / 0	+3.1 / +2.5	1.4 / 4.0	+1.6 / 0	+4.0 / +3.0
1.58–1.97	0.4 / 1.4	+0.6 / 0	+1.4 / +1.0	0.8 / 2.4	+1.0 / 0	+2.4 / +1.8	1.2 / 2.8	+1.0 / 0	+2.8 / +2.2	1.8 / 3.4	+1.0 / 0	+3.4 / +2.8	2.4 / 5.0	+1.6 / 0	+5.0 / +4.0
1.97–2.56	0.6 / 1.8	+0.7 / 0	+1.8 / +1.3	0.8 / 2.7	+1.2 / 0	+2.7 / +2.0	1.3 / 3.2	+1.2 / 0	+3.2 / +2.5	2.3 / 4.2	+1.2 / 0	+4.2 / +3.5	3.2 / 6.2	+1.8 / 0	+6.2 / +5.0
2.56–3.15	0.7 / 1.9	+0.7 / 0	+1.9 / +1.4	1.0 / 2.9	+1.2 / 0	+2.9 / +2.2	1.8 / 3.7	+1.2 / 0	+3.7 / +3.0	2.8 / 4.7	+1.2 / 0	+4.7 / +4.0	4.2 / 7.2	+1.8 / 0	+7.2 / +6.0
3.15–3.94	0.9 / 2.4	+0.9 / 0	+2.4 / +1.8	1.4 / 3.7	+1.4 / 0	+3.7 / +2.8	2.1 / 4.4	+1.4 / 0	+4.4 / +3.5	3.6 / 5.9	+1.4 / 0	+5.9 / +5.0	4.8 / 8.4	+2.2 / 0	+8.4 / +7.0
3.94–4.73	1.1 / 2.6	+0.9 / 0	+2.6 / +2.0	1.6 / 3.9	+1.4 / 0	+3.9 / +3.0	2.6 / 4.9	+1.4 / 0	+4.9 / +4.0	4.6 / 6.9	+1.4 / 0	+6.9 / +6.0	5.8 / 9.4	+2.2 / 0	+9.4 / +8.0
4.73–5.52	1.2 / 2.9	+1.0 / 0	+2.9 / +2.2	1.9 / 4.5	+1.6 / 0	+4.5 / +3.5	3.4 / 6.0	+1.6 / 0	+6.0 / +5.0	5.4 / 8.0	+1.6 / 0	+8.0 / +7.0	7.5 / 11.6	+2.5 / 0	+11.6 / +10.0
5.52–6.30	1.5 / 3.2	+1.0 / 0	+3.2 / +2.5	2.4 / 5.0	+1.6 / 0	+5.0 / +4.0	3.4 / 6.0	+1.6 / 0	+6.0 / +5.0	5.4 / 8.0	+1.6 / 0	+8.0 / +7.0	9.5 / 13.6	+2.5 / 0	+13.6 / +12.0
6.30–7.09	1.8 / 3.5	+1.0 / 0	+3.5 / +2.8	2.9 / 5.5	+1.6 / 0	+5.5 / +4.5	4.4 / 7.0	+1.6 / 0	+7.0 / +6.0	6.4 / 9.0	+1.6 / 0	+9.0 / +8.0	9.5 / 13.6	+2.5 / 0	+13.6 / +12.0
7.09–7.88	1.8 / 3.8	+1.2 / 0	+3.8 / +3.0	3.2 / 6.2	+1.8 / 0	+6.2 / +5.0	5.2 / 8.2	+1.8 / 0	+8.2 / +7.0	7.2 / 10.2	+1.8 / 0	+10.2 / +9.0	11.2 / 15.8	+2.8 / 0	+15.8 / +14.0
7.88–8.86	2.3 / 4.3	+1.2 / 0	+4.3 / +3.5	3.2 / 6.2	+1.8 / 0	+6.2 / +5.0	5.2 / 8.2	+1.8 / 0	+8.2 / +7.0	8.2 / 11.2	+1.8 / 0	+11.2 / +10.0	13.2 / 17.8	+2.8 / 0	+17.8 / +16.0
8.86–9.85	2.3 / 4.3	+1.2 / 0	+4.3 / +3.5	4.2 / 7.2	+1.8 / 0	+7.2 / +6.0	6.2 / 9.2	+1.8 / 0	+9.2 / +8.0	10.2 / 13.2	+1.8 / 0	+13.2 / +12.0	13.2 / 17.8	+2.8 / 0	+17.8 / +16.0
9.85–11.03	2.8 / 4.9	+1.2 / 0	+4.9 / +4.0	4.0 / 7.2	+2.0 / 0	+7.2 / +6.0	7.0 / 10.2	+2.0 / 0	+10.2 / +9.0	10.0 / 13.2	+2.0 / 0	+13.2 / +12.0	15.0 / 20.0	+3.0 / 0	+20.0 / +18.0
11.03–12.41	2.8 / 4.9	+1.2 / 0	+4.9 / +4.0	5.0 / 8.2	+2.0 / 0	+8.2 / +7.0	7.0 / 10.2	+2.0 / 0	+10.2 / +9.0	12.0 / 15.2	+2.0 / 0	+15.2 / +14.0	17.0 / 22.0	+3.0 / 0	+22.0 / +20.0
12.41–13.98	3.1 / 5.5	+1.4 / 0	+5.5 / +4.5	5.8 / 9.4	+2.2 / 0	+9.4 / +8.0	7.8 / 11.4	+2.2 / 0	+11.4 / +10.0	13.8 / 17.4	+2.2 / 0	+17.4 / +16.0	18.5 / 24.2	+3.5 / 0	+24.2 / +22.0
13.98–15.75	3.6 / 6.1	+1.4 / 0	+6.1 / +5.0	5.8 / 9.4	+2.2 / 0	+9.4 / +8.0	9.8 / 13.4	+2.2 / 0	+13.4 / +12.0	15.8 / 19.4	+2.2 / 0	+19.4 / +18.0	21.5 / 27.2	+3.5 / 0	+27.2 / +25.0
15.75–17.72	4.4 / 7.0	+1.6 / 0	+7.0 / +6.0	6.5 / 10.6	+2.5 / 0	+10.6 / +9.0	9.5 / 13.6	+2.5 / 0	+13.6 / +12.0	17.5 / 21.6	+2.5 / 0	+21.6 / +20.0	24.0 / 30.5	+4.0 / 0	+30.5 / +28.0
17.72–19.69	4.4 / 7.0	+1.6 / 0	+7.0 / +6.0	7.5 / 11.6	+2.5 / 0	+11.6 / +10.0	11.5 / 15.6	+2.5 / 0	+15.6 / +14.0	19.5 / 23.6	+2.5 / 0	+23.6 / +22.0	26.0 / 32.5	+4.0 / 0	+32.5 / +30.0

See footnotes at end of table.

All data above heavy lines are in accordance with American-British-Canadian (ABC) agreements. Symbols H6, H7, s6, etc. are hole and shaft designations in ABC system. Limits for sizes above 19.69 inches are not covered by ABC agreements but are given in the ANSI standard.
* Pairs of values shown represent minimum and maximum amounts of interference resulting from application of standard tolerance limits.
Source: Reprinted courtesy of The American Society of Mechanical Engineers.

Table 9. ANSI Interference Locational Fits (LN)

ANSI Standard Interference Locational Fits (ANSI B4.1-1967, R1979)

Tolerance limits given in body of table are added or subtracted to basic size
(as indicated by + or − sign) to obtain maximum and minimum sizes of mating parts.

Nominal Size Range, Inches Over To	Class LN 1			Class LN 2			Class LN 3		
	Limits of Inter-ference	Standard Limits		Limits of Inter-ference	Standard Limits		Limits of Inter-ference	Standard Limits	
		Hole H6	Shaft n5		Hole H7	Shaft p6		Hole H7	Shaft r6
	Values shown below are given in thousandths of an inch								
0–0.12	0 / 0.45	+0.25 / 0	+0.45 / +0.25	0 / 0.65	+0.4 / 0	+0.65 / +0.4	0.1 / 0.75	+0.4 / 0	+0.75 / +0.5
0.12–0.24	0 / 0.5	+0.3 / 0	+0.5 / +0.3	0 / 0.8	+0.5 / 0	+0.8 / +0.5	0.1 / 0.9	+0.5 / 0	+0.9 / +0.6
0.24–0.40	0 / 0.65	+0.4 / 0	+0.65 / +0.4	0 / 1.0	+0.6 / 0	+1.0 / +0.6	0.2 / 1.2	+0.6 / 0	+1.2 / +0.8
0.40–0.71	0 / 0.8	+0.4 / 0	+0.8 / +0.4	0 / 1.1	+0.7 / 0	+1.1 / +0.7	0.3 / 1.4	+0.7 / 0	+1.4 / +1.0
0.71–1.19	0 / 1.0	+0.5 / 0	+1.0 / +0.5	0 / 1.3	+0.8 / 0	+1.3 / +0.8	0.4 / 1.7	+0.8 / 0	+1.7 / +1.2
1.19–1.97	0 / 1.1	+0.6 / 0	+1.1 / +0.6	0 / 1.6	+1.0 / 0	+1.6 / +1.0	0.4 / 2.0	+1.0 / 0	+2.0 / +1.4
1.97–3.15	0.1 / 1.3	+0.7 / 0	+1.3 / +0.8	0.2 / 2.1	+1.2 / 0	+2.1 / +1.4	0.4 / 2.3	+1.2 / 0	+2.3 / +1.6
3.15–4.73	0.1 / 1.6	+0.9 / 0	+1.6 / +1.0	0.2 / 2.5	+1.4 / 0	+2.5 / +1.6	0.6 / 2.9	+1.4 / 0	+2.9 / +2.0
4.73–7.09	0.2 / 1.9	+1.0 / 0	+1.9 / +1.2	0.2 / 2.8	+1.6 / 0	+2.8 / +1.8	0.9 / 3.5	+1.6 / 0	+3.5 / +2.5
7.09–9.85	0.2 / 2.2	+1.2 / 0	+2.2 / +1.4	0.2 / 3.2	+1.8 / 0	+3.2 / +2.0	1.2 / 4.2	+1.8 / 0	+4.2 / +3.0
9.85–12.41	0.2 / 2.3	+1.2 / 0	+2.3 / +1.4	0.2 / 3.4	+2.0 / 0	+3.4 / +2.2	1.5 / 4.7	+2.0 / 0	+4.7 / +3.5
12.41–15.75	0.2 / 2.6	+1.4 / 0	+2.6 / +1.6	0.3 / 3.9	+2.2 / 0	+3.9 / +2.5	2.3 / 5.9	+2.2 / 0	+5.9 / +4.5
15.75–19.69	0.2 / 2.8	+1.6 / 0	+2.8 / +1.8	0.3 / 4.4	+2.5 / 0	+4.4 / +2.8	2.5 / 6.6	+2.5 / 0	+6.6 / +5.0

All data in this table are in accordance with American-British-Canadian (ABC) agreements. Limits for sizes above 19.69 inches are not covered by ABC agreements but are given in the ANSI Standard.
Symbols H7, p6, etc. are hole and shaft designations in ABC system.
*Pairs of values shown represent minimum and maximum amounts of interference resulting from application of standard tolerance limits.
Source: Reprinted courtesy of The American Society of Mechanical Engineers.

Table 10. ANSI Transitional Locational Fits (LT)

ANSI Standard Transitional Locational Fits (ANSI B4.1-1967, R1979)

Nominal Size Range Inches Over To	Class LT 1 Fit*	Class LT 1 Std. Tolerance Limits Hole H7	Class LT 1 Std. Tolerance Limits Shaft js6	Class LT 2 Fit*	Class LT 2 Std. Tolerance Limits Hole H8	Class LT 2 Std. Tolerance Limits Shaft js7	Class LT 3 Fit*	Class LT 3 Std. Tolerance Limits Hole H7	Class LT 3 Std. Tolerance Limits Shaft k6	Class LT 4 Fit*	Class LT 4 Std. Tolerance Limits Hole H8	Class LT 4 Std. Tolerance Limits Shaft k7	Class LT 5 Fit*	Class LT 5 Std. Tolerance Limits Hole H7	Class LT 5 Std. Tolerance Limits Shaft n6	Class LT 6 Fit*	Class LT 6 Std. Tolerance Limits Hole H7	Class LT 6 Std. Tolerance Limits Shaft n7
							Values shown below are in thousandths of an inch											
0–0.12	−0.12 +0.52	+0.4 0	+0.12 −0.12	−0.2 +0.8	+0.6 0	+0.2 −0.2							−0.5 +0.15	+0.4 0	+0.5 +0.25	−0.65 +0.15	+0.4 0	+0.65 +0.25
0.12–0.24	−0.15 +0.65	+0.5 0	+0.15 −0.15	−0.25 +0.95	+0.7 0	+0.25 −0.25							−0.6 +0.2	+0.5 0	+0.6 +0.3	−0.8 +0.2	+0.5 0	+0.8 +0.3
0.24–0.40	−0.2 +0.8	+0.6 0	+0.2 −0.2	−0.3 +1.2	+0.9 0	+0.3 −0.3	−0.5 +0.5	+0.6 0	+0.5 +0.1	−0.7 +0.8	+0.9 0	+0.7 +0.1	−0.8 +0.2	+0.6 0	+0.8 +0.4	−1.0 +0.2	+0.6 0	+1.0 +0.4
0.40–0.71	−0.2 +0.9	+0.7 0	+0.2 −0.2	−0.35 +1.35	+1.0 0	+0.35 −0.35	−0.5 +0.6	+0.7 0	+0.5 +0.1	−0.8 +0.9	+1.0 0	+0.8 +0.1	−0.9 +0.2	+0.7 0	+0.9 +0.5	−1.2 +0.2	+0.7 0	+1.2 +0.5
0.71–1.19	−0.25 +1.05	+0.8 0	+0.25 −0.25	−0.4 +1.6	+1.2 0	+0.4 −0.4	−0.6 +0.7	+0.8 0	+0.6 +0.1	−0.9 +1.1	+1.2 0	+0.9 +0.1	−1.1 +0.2	+0.8 0	+1.1 +0.6	−1.4 +0.2	+0.8 0	+1.4 +0.6
1.19–1.97	−0.3 +1.3	+1.0 0	+0.3 −0.3	−0.5 +2.1	+1.6 0	+0.5 −0.5	−0.7 +0.9	+1.0 0	+0.7 +0.1	−1.1 +1.5	+1.6 0	+1.1 +0.1	−1.3 +0.3	+1.0 0	+1.3 +0.7	−1.7 +0.3	+1.0 0	+1.7 +0.7
1.97–3.15	−0.3 +1.5	+1.2 0	+0.3 −0.3	−0.6 +2.4	+1.8 0	+0.6 −0.6	−0.8 +1.1	+1.2 0	+0.8 +0.1	−1.3 +1.7	+1.8 0	+1.3 +0.1	−1.5 +0.4	+1.2 0	+1.5 +0.8	−2.0 +0.4	+1.2 0	+2.0 +0.8
3.15–4.73	−0.4 +1.8	+1.4 0	+0.4 −0.4	−0.7 +2.9	+2.2 0	+0.7 −0.7	−1.0 +1.3	+1.4 0	+1.0 +0.1	−1.5 +2.1	+2.2 0	+1.5 +0.1	−1.9 +0.4	+1.4 0	+1.9 +1.0	−2.4 +0.4	+1.4 0	+2.4 +1.0
4.73–7.09	−0.5 +2.1	+1.6 0	+0.5 −0.5	−0.8 +3.3	+2.5 0	+0.8 −0.8	−1.1 +1.5	+1.6 0	+1.1 +0.1	−1.7 +2.4	+2.5 0	+1.7 +0.1	−2.2 +0.4	+1.6 0	+2.2 +1.2	−2.8 +0.4	+1.6 0	+2.8 +1.2
7.09–9.85	−0.6 +2.4	+1.8 0	+0.6 −0.6	−0.9 +3.7	+2.8 0	+0.9 −0.9	−1.4 +1.6	+1.8 0	+1.4 +0.2	−2.0 +2.6	+2.8 0	+2.0 +0.2	−2.6 +0.4	+1.8 0	+2.6 +1.4	−3.2 +0.4	+1.8 0	+3.2 +1.4
9.85–12.41	−0.6 +2.6	+2.0 0	+0.6 −0.6	−1.0 +4.0	+3.0 0	+1.0 −1.0	−1.4 +1.8	+2.0 0	+1.4 +0.2	−2.2 +2.8	+3.0 0	+2.2 +0.2	−2.6 +0.6	+2.0 0	+2.6 +1.4	−3.4 +0.6	+2.0 0	+3.4 +1.4
12.41–15.75	−0.7 +2.9	+2.2 0	+0.7 −0.7	−1.0 +4.5	+3.5 0	+1.0 −1.0	−1.6 +2.0	+2.2 0	+1.6 +0.2	−2.4 +3.3	+3.5 0	+2.4 +0.2	−3.0 +0.6	+2.2 0	+3.0 +1.6	−3.8 +0.6	+2.2 0	+3.8 +1.6
15.75–19.69	−0.8 +3.3	+2.5 0	+0.8 −0.8	−1.2 +5.2	+4.0 0	+1.2 −1.2	−1.8 +2.3	+2.5 0	+1.8 +0.2	−2.7 +3.8	+4.0 0	+2.7 +0.2	−3.4 +0.7	+2.5 0	+3.4 +1.8	−4.3 +0.7	+2.5 0	+4.3 +1.8

All data above heavy lines are in accord with ABC agreements. Symbols H7, js6. etc. are hole and shaft designations in ABC system.
* Pairs of values shown represent maximum amount of interference (−) and maximum amount of clearance (+) resulting from application of standard tolerance limits.
Source: Reprinted courtesy of The American Society of Mechanical Engineers.

Appendix A   199

Table 11. ANSI Clearance Locational Fits (LC)

ANSI National Standard Clearance Locational Fits (ANSI B4.1-1967, R1979)

Tolerance limits given in body of table are added or subtracted to basic size (as indicated by + or – sign) to obtain maximum and minimum sizes of mating parts.

Nominal Size Range, Inches Over To	Class LC 1			Class LC 2			Class LC 3			Class LC 4			Class LC 5		
	Clearance*	Standard Tolerance Limits		Clearance*	Standard Tolerance Limits		Clearance*	Standard Tolerance Limits		Clearance*	Standard Tolerance Limits		Clearance*	Standard Tolerance Limits	
		Hole H6	Shaft h5		Hole H7	Shaft h6		Hole H8	Shaft h7		Hole H10	Shaft h9		Hole H7	Shaft g6
	Values shown below are in thousandths of an inch														
0–0.12	0 / 0.45	+0.25 / 0	0 / −0.2	0 / 0.65	+0.4 / 0	0 / −0.25	0 / 1	+0.6 / 0	0 / −0.4	0 / 2.6	+1.6 / 0	0 / −1.0	0.1 / 0.75	+0.4 / 0	−0.1 / −0.35
0.12–0.24	0 / 0.5	+0.3 / 0	0 / −0.2	0 / 0.8	+0.5 / 0	0 / −0.3	0 / 1.2	+0.7 / 0	0 / −0.5	0 / 3.0	+1.8 / 0	0 / −1.2	0.15 / 0.95	+0.5 / 0	−0.15 / −0.45
0.24–0.40	0 / 0.65	+0.4 / 0	0 / −0.25	0 / 1.0	+0.6 / 0	0 / −0.4	0 / 1.5	+0.9 / 0	0 / −0.6	0 / 3.6	+2.2 / 0	0 / −1.4	0.2 / 1.2	+0.6 / 0	−0.2 / −0.6
0.40–0.71	0 / 0.7	+0.4 / 0	0 / −0.3	0 / 1.1	+0.7 / 0	0 / −0.4	0 / 1.7	+1.0 / 0	0 / −0.7	0 / 4.4	+2.8 / 0	0 / −1.6	0.25 / 1.35	+0.7 / 0	−0.25 / −0.65
0.71–1.19	0 / 0.9	+0.5 / 0	0 / −0.4	0 / 1.3	+0.8 / 0	0 / −0.5	0 / 2	+1.2 / 0	0 / −0.8	0 / 5.5	+3.5 / 0	0 / −2.0	0.3 / 1.6	+0.8 / 0	−0.3 / −0.8
1.19–1.97	0 / 1.0	+0.6 / 0	0 / −0.4	0 / 1.6	+1.0 / 0	0 / −0.6	0 / 2.6	+1.6 / 0	0 / −1	0 / 6.5	+4.0 / 0	0 / −2.5	0.4 / 2.0	+1.0 / 0	−0.4 / −1.0
1.97–3.15	0 / 1.2	+0.7 / 0	0 / −0.5	0 / 1.9	+1.2 / 0	0 / −0.7	0 / 3	+1.8 / 0	0 / −1.2	0 / 7.5	+4.5 / 0	0 / −3	0.4 / 2.3	+1.2 / 0	−0.4 / −1.1
3.15–4.73	0 / 1.5	+0.9 / 0	0 / −0.6	0 / 2.3	+1.4 / 0	0 / −0.9	0 / 3.6	+2.2 / 0	0 / −1.4	0 / 8.5	+5.0 / 0	0 / −3.5	0.5 / 2.8	+1.4 / 0	−0.5 / −1.4
4.73–7.09	0 / 1.7	+1.0 / 0	0 / −0.7	0 / 2.6	+1.6 / 0	0 / −1.0	0 / 4.1	+2.5 / 0	0 / −1.6	0 / 10.0	+6.0 / 0	0 / −4	0.6 / 3.2	+1.6 / 0	−0.6 / −1.6
7.09–9.85	0 / 2.0	+1.2 / 0	0 / −0.8	0 / 3.0	+1.8 / 0	0 / −1.2	0 / 4.6	+2.8 / 0	0 / −1.8	0 / 11.5	+7.0 / 0	0 / −4.5	0.6 / 3.6	+1.8 / 0	−0.6 / −1.8
9.85–12.41	0 / 2.1	+1.2 / 0	0 / −0.9	0 / 3.2	+2.0 / 0	0 / −1.2	0 / 5	+3.0 / 0	0 / −2.0	0 / 13.0	+8.0 / 0	0 / −5	0.7 / 3.9	+2.0 / 0	−0.7 / −1.9
12.41–15.75	0 / 2.4	+1.4 / 0	0 / −1.0	0 / 3.6	+2.2 / 0	0 / −1.4	0 / 5.7	+3.5 / 0	0 / −2.2	0 / 15.0	+9.0 / 0	0 / −6	0.7 / 4.3	+2.2 / 0	−0.7 / −2.1
15.75–19.69	0 / 2.6	+1.6 / 0	0 / −1.0	0 / 4.1	+2.5 / 0	0 / −1.6	0 / 6.5	+4 / 0	0 / −2.5	0 / 16.0	+10.0 / 0	0 / −6	0.8 / 4.9	+2.5 / 0	−0.8 / −2.4

See footnotes at end of table.

Nominal Size Range Inches Over To	Class LC 6			Class LC 7			Class LC 8			Class LC 9			Class LC 10			Class LC 11		
	Clearance*	Std. Tolerance Limits		Clearance*	Std. Tolerance Limits		Clearance*	Std. Tolerance Limits		Clearance*	Std. Tolerance Limits		Clearance*	Std. Tolerance Limits		Clearance*	Std. Tolerance Limits	
		Hole H9	Shaft f8		Hole H10	Shaft e9		Hole H10	Shaft d9		Hole H11	Shaft c10		Hole H12	Shaft		Hole H13	Shaft
	Values shown below are in thousandths of an inch																	
0–0.12	0.3 / 1.9	+1.0 / 0	−0.3 / −0.9	0.6 / 3.2	+1.6 / 0	−0.6 / −1.6	1.0 / 2.0	+1.6 / 0	−1.0 / −2.0	2.5 / 6.6	+2.5 / 0	−2.5 / −4.1	4 / 12	+4 / 0	−4 / −8	5 / 17	+6 / 0	−5 / −11
0.12–0.24	0.4 / 2.3	+1.2 / 0	−0.4 / −1.1	0.8 / 3.8	+1.8 / 0	−0.8 / −2.0	1.2 / 4.2	+1.8 / 0	−1.2 / −2.4	2.8 / 7.6	+3.0 / 0	−2.8 / −4.6	4.5 / 14.5	+5 / 0	−4.5 / −9.5	6 / 20	+7 / 0	−6 / −13
0.24–0.40	0.5 / 2.8	+1.4 / 0	−0.5 / −1.4	1.0 / 4.6	+2.2 / 0	−1.0 / −2.4	1.6 / 5.2	+2.2 / 0	−1.6 / −3.0	3.0 / 8.7	+3.5 / 0	−3.0 / −5.2	5 / 17	+6 / 0	−5 / −11	7 / 25	+9 / 0	−7 / −16
0.40–0.71	0.6 / 3.2	+1.6 / 0	−0.6 / −1.6	1.2 / 5.6	+2.8 / 0	−1.2 / −2.8	2.0 / 6.4	+2.8 / 0	−2.0 / −3.6	3.5 / 10.3	+4.0 / 0	−3.5 / −6.3	6 / 20	+7 / 0	−6 / −13	8 / 28	+10 / 0	−8 / −18
0.71–1.19	0.8 / 4.0	+2.0 / 0	−0.8 / −2.0	1.6 / 7.1	+3.5 / 0	−1.6 / −3.6	2.5 / 8.0	+3.5 / 0	−2.5 / −4.5	4.5 / 13.0	+5.0 / 0	−4.5 / −8.0	7 / 23	+8 / 0	−7 / −15	10 / 34	+12 / 0	−10 / −22
1.19–1.97	1.0 / 5.1	+2.5 / 0	−1.0 / −2.6	2.0 / 8.5	+4.0 / 0	−2.0 / −4.5	3.6 / 9.5	+4.0 / 0	−3.0 / −5.5	5.0 / 15.0	+6 / 0	−5.0 / −9.0	8 / 28	+10 / 0	−8 / −18	12 / 44	+16 / 0	−12 / −28
1.97–3.15	1.2 / 6.0	+3.0 / 0	−1.0 / −3.0	2.5 / 10.0	+4.5 / 0	−2.5 / −5.5	4.0 / 11.5	+4.5 / 0	−4.0 / −7.0	6.0 / 17.5	+7 / 0	−6.0 / −10.5	10 / 34	+12 / 0	−10 / −22	14 / 50	+18 / 0	−14 / −32
3.15–4.73	1.4 / 7.1	+3.5 / 0	−1.4 / −3.6	3.0 / 11.5	+5.0 / 0	−3.0 / −6.5	5.0 / 13.5	+5.0 / 0	−5.0 / −8.5	7 / 21	+9 / 0	−7 / −12	11 / 39	+14 / 0	−11 / −25	16 / 60	+22 / 0	−16 / −38
4.73–7.09	1.6 / 8.1	+4.0 / 0	−1.6 / −4.1	3.5 / 13.5	+6.0 / 0	−3.5 / −7.5	6 / 16	+6 / 0	−6 / −10	8 / 24	+10 / 0	−8 / −14	12 / 44	+16 / 0	−12 / −28	18 / 68	+25 / 0	−18 / −43
7.09–9.85	2.0 / 9.3	+4.5 / 0	−2.0 / −4.8	4.0 / 15.5	+7.0 / 0	−4.0 / −8.5	7 / 18.5	+7 / 0	−7 / −11.5	10 / 29	+12 / 0	−10 / −17	16 / 52	+18 / 0	−16 / −34	22 / 78	+28 / 0	−22 / −50
9.85–12.41	2.2 / 10.2	+5.0 / 0	−2.2 / −5.2	4.5 / 17.5	+8.0 / 0	−4.5 / −9.5	7 / 20	+8 / 0	−7 / −12	12 / 32	+12 / 0	−12 / −20	20 / 60	+20 / 0	−20 / −40	28 / 88	+30 / 0	−28 / −58
12.41–15.75	2.5 / 12.0	+6.0 / 0	−2.5 / −6.0	5.0 / 20.0	+9.0 / 0	−5 / −11	8 / 23	+9 / 0	−8 / −14	14 / 37	+14 / 0	−14 / −23	22 / 66	+22 / 0	−22 / −44	30 / 100	+35 / 0	−30 / −65
15.75–19.69	2.8 / 12.8	+6.0 / 0	−2.8 / −6.8	5.0 / 21.0	+10.0 / 0	−5 / −11	9 / 25	+10 / 0	−9 / −15	16 / 42	+16 / 0	−16 / −26	25 / 75	+25 / 0	−25 / −50	35 / 115	+40 / 0	−35 / −75

All data above heavy lines are in accordance with American-British-Canadian (ABC) agreements. Symbols H6, H7, s6, etc. are hole and shaft designations in ABC system. Limits for sizes above 19.69 inches are not covered by ABC agreements but are given in the ANSI Standard.
* Pairs of values shown represent minimum and maximum amounts of interference resulting from application of standard tolerance limits.
Source: Reprinted courtesy of The American Society of Mechanical Engineers.

Table 12. ANSI Running and Sliding Fits (RC)

American National Standard Running and Sliding Fits (ANSI B4.1-1967, RI979)

Tolerance limits given in body of table are added or subtracted to basic size (as indicated by + or − sign) to obtain maximum and minimum sizes of mating parts.

Nominal Size Range, Inches Over To	Class RC 1 Clear-ance*	Class RC 1 Standard Tolerance Limits Hole H5	Class RC 1 Standard Tolerance Limits Shaft g4	Class RC 2 Clear-ance*	Class RC 2 Standard Tolerance Limits Hole H6	Class RC 2 Standard Tolerance Limits Shaft g5	Class RC 3 Clear-ance*	Class RC 3 Standard Tolerance Limits Hole H7	Class RC 3 Standard Tolerance Limits Shaft f6	Class RC 4 Clear-ance*	Class RC 4 Standard Tolerance Limits Hole H8	Class RC 4 Standard Tolerance Limits Shaft f7
					Values shown below are in thousandths of an inch							
0–0.12	0.1 / 0.45	+0.2 / 0	−0.1 / −0.25	0.1 / 0.55	+0.25 / 0	−0.1 / −0.3	0.3 / 0.95	+0.4 / 0	−0.3 / −0.55	0.3 / 1.3	+0.6 / 0	−0.3 / −0.7
0.12–0.24	0.15 / 0.5	+0.2 / 0	−0.15 / −0.3	0.15 / 0.65	+0.3 / 0	−0.15 / −0.35	0.4 / 1.12	+0.5 / 0	−0.4 / −0.7	0.4 / 1.6	+0.7 / 0	−0.4 / −0.9
0.24–0.40	0.2 / 0.6	+0.25 / 0	−0.2 / −0.35	0.2 / 0.85	+0.4 / 0	−0.2 / −0.45	0.5 / 1.5	+0.6 / 0	−0.5 / −0.9	0.5 / 2.0	+0.9 / 0	−0.5 / −1.1
0.40–0.71	0.25 / 0.75	+0.3 / 0	−0.25 / −0.45	0.25 / 0.95	+0.4 / 0	−0.25 / −0.55	0.6 / 1.7	+0.7 / 0	−0.6 / −1.0	0.6 / 2.3	+1.0 / 0	−0.6 / −1.3
0.71–1.19	0.3 / 0.95	+0.4 / 0	−0.3 / −0.55	0.3 / 1.2	+0.5 / 0	−0.3 / −0.7	0.8 / 2.1	+0.8 / 0	−0.8 / −1.3	0.8 / 2.8	+1.2 / 0	−0.8 / −1.6
1.19–1.97	0.4 / 1.1	+0.4 / 0	−0.4 / −0.7	0.4 / 1.4	+0.6 / 0	−0.4 / −0.8	1.0 / 2.6	+1.0 / 0	−1.0 / −1.6	1.0 / 3.6	+1.6 / 0	−1.0 / −2.0
1.97–3.15	0.4 / 1.2	+0.5 / 0	−0.4 / −0.7	0.4 / 1.6	+0.7 / 0	−0.4 / −0.9	1.2 / 3.1	+1.2 / 0	−1.2 / −1.9	1.2 / 4.2	+1.8 / 0	−1.2 / −2.4
3.15–4.73	0.5 / 1.5	+0.6 / 0	−0.5 / −0.9	0.5 / 2.0	+0.9 / 0	−0.5 / −1.1	1.4 / 3.7	+1.4 / 0	−1.4 / −2.3	1.4 / 5.0	+2.2 / 0	−1.4 / −2.8
4.73–7.09	0.6 / 1.8	+0.7 / 0	−0.6 / −1.1	0.6 / 2.3	+1.0 / 0	−0.6 / −1.3	1.6 / 4.2	+1.6 / 0	−1.6 / −2.6	1.6 / 5.7	+2.5 / 0	−1.6 / −3.2
7.09–9.85	0.6 / 2.0	+0.8 / 0	−0.6 / −1.2	0.6 / 2.6	+1.2 / 0	−0.6 / −1.4	2.0 / 5.0	+1.8 / 0	−2.0 / −3.2	2.0 / 6.6	+2.8 / 0	−2.0 / −3.8
9.85–12.41	0.8 / 2.3	+0.9 / 0	−0.8 / −1.4	0.8 / 2.9	+1.2 / 0	−0.8 / −1.7	2.5 / 5.7	+2.0 / 0	−2.5 / −3.7	2.5 / 7.5	+3.0 / 0	−2.5 / −4.5
12.41–15.75	1.0 / 2.7	+1.0 / 0	−1.0 / −1.7	1.0 / 3.4	+1.4 / 0	−1.0 / −2.0	3.0 / 6.6	+2.2 / 0	−3.0 / −4.4	3.0 / 8.7	+3.5 / 0	−3.0 / −5.2
15.75–19.69	1.2 / 3.0	+1.0 / 0	−1.2 / −2.0	1.2 / 3.8	+1.6 / 0	−1.2 / −2.2	4.0 / 8.1	+2.5 / 0	−4.0 / −5.6	4.0 / 10.5	+4.0 / 0	−4.0 / −6.5

See footnotes at end of table.

Nominal Size Range, Inches Over To	Class RC 5 Clear-ance*	Class RC 5 Standard Tolerance Limits Hole H8	Class RC 5 Standard Tolerance Limits Shaft e7	Class RC 6 Clear-ance*	Class RC 6 Standard Tolerance Limits Hole H9	Class RC 6 Standard Tolerance Limits Shaft e8	Class RC 7 Clear-ance*	Class RC 7 Standard Tolerance Limits Hole H9	Class RC 7 Standard Tolerance Limits Shaft d8	Class RC 8 Clear-ance*	Class RC 8 Standard Tolerance Limits Hole H10	Class RC 8 Standard Tolerance Limits Shaft c9	Class RC 9 Clear-ance*	Class RC 9 Standard Tolerance Limits Hole H11	Class RC 9 Standard Tolerance Limits Shaft
						Values shown below are in thousandths of an inch									
0–0.12	0.6 / 1.6	+0.6 / 0	−0.6 / −1.0	0.6 / 2.2	+1.0 / 0	−0.6 / −1.2	1.0 / 2.6	+1.0 / 0	−1.0 / −1.6	2.5 / 5.1	+1.6 / 0	−2.5 / −3.5	4.0 / 8.1	+2.5 / 0	−4.0 / −5.6
0.12–0.24	0.8 / 2.0	+0.7 / 0	−0.8 / −1.3	0.8 / 2.7	+1.2 / 0	−0.8 / −1.5	1.2 / 3.1	+1.2 / 0	−1.2 / −1.9	2.8 / 5.8	+1.8 / 0	−2.8 / −4.0	4.5 / 9.0	+3.0 / 0	−4.5 / −6.0
0.24–0.40	1.0 / 2.5	+0.9 / 0	−1.0 / −1.6	1.0 / 3.3	+1.4 / 0	−1.0 / −1.9	1.6 / 3.9	+1.4 / 0	−1.6 / −2.5	3.0 / 6.6	+2.2 / 0	−3.0 / −4.4	5.0 / 10.7	+3.5 / 0	−5.0 / −7.2
0.40–0.71	1.2 / 2.9	+1.0 / 0	−1.2 / −1.9	1.2 / 3.8	+1.6 / 0	−1.2 / −2.2	2.0 / 4.6	+1.6 / 0	−2.0 / −3.0	3.5 / 7.9	+2.8 / 0	−3.5 / −5.1	6.0 / 12.8	+4.0 / 0	−6.0 / −8.8
0.71–1.19	1.6 / 3.6	+1.2 / 0	−1.6 / −2.4	1.6 / 4.8	+2.0 / 0	−1.6 / −2.8	2.5 / 5.7	+2.0 / 0	−2.5 / −3.7	4.5 / 10.0	+3.5 / 0	−4.5 / −6.5	7.0 / 15.5	+5.0 / 0	−7.0 / −10.5
1.19–1.97	2.0 / 4.6	+1.6 / 0	−2.0 / −3.0	2.0 / 6.1	+2.5 / 0	−2.0 / −3.6	3.0 / 7.1	+2.5 / 0	−3.0 / −4.6	5.0 / 11.5	+4.0 / 0	−5.0 / −7.5	8.0 / 18.0	+6.0 / 0	−8.0 / −12.0
1.97–3.15	2.5 / 5.5	+1.8 / 0	−2.5 / −3.7	2.5 / 7.3	+3.0 / 0	−2.5 / −4.3	4.0 / 8.8	+3.0 / 0	−4.0 / −5.8	6.0 / 13.5	+4.5 / 0	−6.0 / −9.0	9.0 / 20.5	+7.0 / 0	−9.0 / −13.5
3.15–4.73	3.0 / 6.6	+2.2 / 0	−3.0 / −4.4	3.0 / 8.7	+3.5 / 0	−3.0 / −5.2	5.0 / 10.7	+3.5 / 0	−5.0 / −7.2	7.0 / 15.5	+5.0 / 0	−7.0 / −10.5	10.0 / 24.0	+9.0 / 0	−10.0 / −15.0
4.73–7.09	3.5 / 7.6	+2.5 / 0	−3.5 / −5.1	3.5 / 10.0	+4.0 / 0	−3.5 / −6.0	6.0 / 12.5	+4.0 / 0	−6.0 / −8.5	8.0 / 18.0	+6.0 / 0	−8.0 / −12.0	12.0 / 28.0	+10.0 / 0	−12.0 / −18.0
7.09–9.85	4.0 / 8.6	+2.8 / 0	−4.0 / −5.8	4.0 / 11.3	+4.5 / 0	−4.0 / −6.8	7.0 / 14.3	+4.5 / 0	−7.0 / −9.8	10.0 / 21.5	+7.0 / 0	−10.0 / −14.5	15.0 / 34.0	+12.0 / 0	−15.0 / −22.0
9.85–12.41	5.0 / 10.0	+3.0 / 0	−5.0 / −7.0	5.0 / 13.0	+5.0 / 0	−5.0 / −8.0	8.0 / 16.0	+5.0 / 0	−8.0 / −11.0	12.0 / 25.0	+8.0 / 0	−12.0 / −17.0	18.0 / 38.0	+12.0 / 0	−18.0 / −26.0
12.41–15.75	6.0 / 11.7	+3.5 / 0	−6.0 / −8.2	6.0 / 15.5	+6.0 / 0	−6.0 / −9.5	10.0 / 19.5	+6.0 / 0	−10.0 / −13.5	14.0 / 29.0	+9.0 / 0	−14.0 / −20.0	22.0 / 45.0	+14.0 / 0	−22.0 / −31.0
15.75–19.69	8.0 / 14.5	+4.0 / 0	−8.0 / −10.5	8.0 / 18.0	+6.0 / 0	−8.0 / −12.0	12.0 / 22.0	+6.0 / 0	−12.0 / −16.0	16.0 / 32.0	+10.0 / 0	−16.0 / −22.0	25.0 / 51.0	+16.0 / 0	−25.0 / −35.0

All data above heavy lines are in accord with ABC agreements. Symbols H5, g4, etc. are hole and shaft designations in ABC system. Limits for sizes above 19.69 inches are also given in the ANSI Standard.

* Pairs of values shown represent minimum and maximum amounts of clearance resulting from application of standard tolerance limits.

Source: Reprinted courtesy of The American Society of Mechanical Engineers.

Table 13. Square Head Bolts

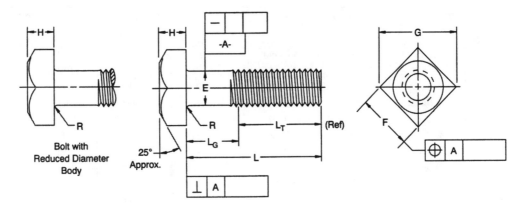

Nominal Size or Basic Product Diameter		E Body Diameter	F Width Across Flats			G Width Across Corners		H Height			R Radius of Fillet		L_T Thread Length for Bolt Lengths	
													6 in. and shorter	over 6 in.
		Max	Basic	Max	Min	Max	Min	Basic	Max	Min	Max	Min	Basic	Basic
1/4	0.2500	0.260	3/8	0.375	0.362	0.530	0.498	11/64	0.188	0.156	0.03	0.01	0.750	1.000
5/16	0.3125	0.324	1/2	0.500	0.484	0.707	0.665	13/64	0.220	0.186	0.03	0.01	0.875	1.125
3/8	0.3750	0.388	9/16	0.562	0.544	0.795	0.747	1/4	0.268	0.232	0.03	0.01	1.000	1.250
7/16	0.4375	0.452	5/8	0.625	0.603	0.884	0.828	19/64	0.316	0.278	0.03	0.01	1.125	1.375
1/2	0.5000	0.515	3/4	0.750	0.725	1.061	0.995	21/64	0.348	0.308	0.03	0.01	1.250	1.500
5/8	0.6250	0.642	15/16	0.938	0.906	1.326	1.244	27/64	0.444	0.400	0.06	0.02	1.500	1.750
3/4	0.7500	0.768	1-1/8	1.125	1.088	1.591	1.494	1/2	0.524	0.476	0.06	0.02	1.750	2.000
7/8	0.8750	0.895	1-5/16	1.312	1.269	1.856	1.742	19/32	0.620	0.568	0.06	0.02	2.000	2.250
1	1.0000	1.022	1-1/2	1.500	1.450	2.121	1.991	21/32	0.684	0.628	0.09	0.03	2.250	2.500
1-1/8	1.1250	1.149	1-11/16	1.688	1.631	2.386	2.239	3/4	0.780	0.720	0.09	0.03	2.500	2.750
1-1/4	1.2500	1.277	1-7/8	1.875	1.812	2.652	2.489	27/32	0.876	0.812	0.09	0.03	2.750	3.000
1-3/8	1.3750	1.404	2-1/16	2.062	1.994	2.917	2.738	29/32	0.940	0.872	0.09	0.03	3.000	3.250
1-1/2	1.5000	1.531	2-1/4	2.250	2.175	3.182	2.986	1	1.036	0.964	0.09	0.03	3.250	3.500

Source: Reprinted courtesy of The American Society of Mechanical Engineers.

Table 14. Tap Drill Sizes for American National Thread Forms

Screw Thread		Commercial Tap Drills*		Screw Thread		Commercial Tap Drills*	
Outside Diam. Pitch	Root Diam.	Size or Number	Decimal Equiv.	Outside Diam. Pitch	Root Diam.	Size or Number	Decimal Equiv.
1/16–64	0.0422	3/64	0.0469	27	0.4519	15/32	0.4687
72	0.0445	3/64	0.0469	9/16–12	0.4542	31/64	0.4844
5/64–60	0.0563	1/16	0.0625	18	0.4903	33/64	0.5156
72	0.0601	52	0.0635	27	0.5144	17/32	0.5312
3/32–48	0.0667	49	0.0730	5/8–11	0.5069	17/32	0.5312
50	0.0678	49	0.0730	12	0.5168	35/64	0.5469
7/64–48	0.0823	43	0.0890	18	0.5528	37/64	0.5781
1/8–32	0.0844	3/32	0.0937	27	0.5769	19/32	0.5937
40	0.0925	38	0.1015	11/16–11	0.5694	19/32	0.5937
9/64–40	0.1081	32	0.1160	16	0.6063	5/8	0.6250
5/32–32	0.1157	1/8	0.1250	3/4–10	0.6201	21/32	0.6562
36	0.1202	30	0.1285	12	0.6418	43/64	0.6719
11/64–32	0.1313	9/64	0.1406	16	0.6688	11/16	0.6875
3/16–24	0.1334	26	0.1470	27	0.7019	23/32	0.7187
32	0.1469	22	0.1570	13/16–10	0.6826	23/32	0.7187
13/64–24	0.1490	20	0.1610	7/8–9	0.7307	49/64	0.7656
7/32–24	0.1646	16	0.1770	12	0.7668	51/64	0.7969
32	0.1782	12	0.1890	14	0.7822	13/16	0.8125
15/64–24	0.1806	10	0.1935	18	0.8028	53/64	0.8281
1/4–20	0.1850	7	0.2010	27	0.8269	27/32	0.8437
24	0.1959	4	0.2090	15/16–9	0.7932	53/64	0.8281
27	0.2019	3	0.2130	1–8	0.8376	7/8	0.8750
28	0.2036	3	0.2130	12	0.8918	59/64	0.9219
32	0.2094	7/32	0.2187	14	0.9072	15/16	0.9375
5/16–18	0.2403	F	0.2570	27	0.9519	31/32	0.9687
20	0.2476	17/64	0.2656	1 1/8–7	0.9394	63/64	0.9844
24	0.2584	I	0.2720	12	1.0168	1 3/64	1.0469
27	0.2644	J	0.2770	1 1/4–7	1.0644	1 7/64	1.1094
32	0.2719	9/32	0.2812	12	1.1418	1 11/64	1.1719
3/8–16	0.2938	5/16	0.3125	1 3/8–6	1.1585	1 7/32	1.2187
20	0.3100	21/64	0.3281	12	1.2668	1 19/64	1.2969
24	0.3209	Q	0.3320	1 1/2–6	1.2835	1 11/32	1.3437
27	0.3269	R	0.3390	12	1.3918	1 27/64	1.4219
7/16–14	0.3447	U	0.3680	1 5/6–5 1/2	1.3888	1 20/64	1.4531
20	0.3726	25/64	0.3906	1 3/4–5	1.4902	1 9/16	1.5625
24	0.3834	X	0.3970	1 7/8–5	1.6152	1 11/16	1.6875
27	0.3894	Y	0.4040	2–4 1/2	1.7113	1 25/32	1.7812
1/2–12	0.3918	27/64	0.4219	2 1/8–4 1/2	1.8363	1 29/32	1.9062
13	0.4001	27/64	0.4219	2 1/4–4 1/2	1.9613	2 1/32	2.0312
20	0.4351	29/64	0.4531	2 3/8–4	2.0502	2 1/8	2.1250
24	0.4459	29/64	0.4531	2 1/2–4	2.1752	2 1/4	2.2500

*These tap drill diameters allow approximately 75 percent of a full thread to be produced. For small thread sizes in the first column, the use of drills to produce the larger hole sizes shown in Table 2 will reduce defects caused by tap problems and breakage.

Table 15. Hex Cap Screws (Finished Hex Bolts)

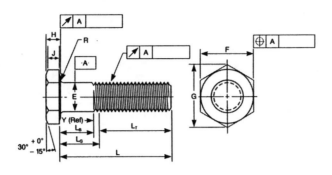

Nominal Size or Basic Product Dia.		E Body Diameter		F Width Across Flats			G Width Across Corners		H Height			J Wrenching Height	L_T Thread Length for Screw Lengths		Y Transition Thread Length	Runout of Bearing Surface FIM
													6 in. and shorter	Over 6 in.		
		Max	Min	Basic	Max	Min	Max	Min	Basic	Max	Min	Min	Basic	Basic	Max	Max
1/4	0.2500	0.2500	0.2450	7/16	0.438	0.428	0.505	0.488	5/32	0.163	0.150	0.106	0.750	1.000	0.250	0.10
5/16	0.3125	0.3125	0.3065	1/2	0.500	0.489	0.577	0.557	13/64	0.211	0.195	0.140	0.875	1.125	0.278	0.011
3/8	0.3750	0.3750	0.3690	9/16	0.562	0.551	0.650	0.628	15/64	0.243	0.226	0.160	1.000	1.250	0.312	0.012
7/16	0.4375	0.4375	0.4305	5/8	0.625	0.612	0.722	0.698	9/32	0.291	0.272	0.195	1.125	1.375	0.357	0.013
1/2	0.5000	0.5000	0.4930	3/4	0.750	0.736	0.866	0.840	5/16	0.323	0.302	0.215	1.250	1.500	0.385	0.074
9/16	0.5625	0.5625	0.5545	13/16	0.812	0.798	0.938	0.910	23/64	0.371	0.348	0.250	1.375	1.625	0.417	0.015
5/8	0.6250	0.6250	0.6170	15/16	0.938	0.922	1.083	1.051	25/64	0.403	0.378	0.269	1.500	1.750	0.455	0.017
3/4	0.7500	0.7500	0.7410	1-1/8	1.125	1.100	1.299	1.254	15/32	0.483	0.455	0.324	1.750	2.000	0.500	0.020
7/8	0.8750	0.8750	0.8660	1-5/16	1.312	1.285	1.516	1.465	35/64	0.563	0.531	0.378	2.000	2.250	0.556	0.023
1	1.0000	1.0000	0.9900	1-1/2	1.500	1.469	1.732	1.675	39/64	0.627	0.591	0.416	2.250	2.500	0.625	0.026
1-1/8	1.1250	1.1250	1.1140	1-11/16	1.688	1.631	1.949	1.859	11/16	0.718	0.658	0.461	2.500	2.750	0.714	0.029
1-1/4	1.2500	1.2500	1.2390	1-7/8	1.875	1.812	2.165	2.066	25/32	0.813	0.749	0.530	2.750	3.000	0.714	0.033
1-3/8	1.3750	1.3750	1.3630	2-1/16	2.062	1.994	2.382	2.273	27/32	0.878	0.810	0.569	3.000	3.250	0.833	0.036
1-1/2	1.5000	1.5000	1.4880	2-1/4	2.250	2.175	2.598	2.480	1-5/16	0.974	0.902	0.640	3.250	3.500	0.833	0.039
1-3/4	1.7500	1.7500	1.7380	2-5/8	2.625	2.538	3.031	2.893	1-3/32	1.134	1.054	0.748	3.750	4.000	1.000	0.046
2	2.0000	2.0000	1.9880	3	3.000	2.900	3.464	3.306	1-7/32	1.263	1.175	0.825	4.250	4.500	1.111	0.052
2-1/4	2.2500	2.2500	2.2380	3-3/8	3.375	3.262	3.897	3.719	1-3/8	1.423	1.327	0.933	4.750	5.000	1.111	0.059
2-1/2	2.5000	2.5000	2.4880	3-3/4	3.750	3.625	4.330	4.133	1-17/32	1.583	1.479	1.042	5.250	5.500	1.250	0.065
2-3/4	2.7500	2.7500	2.7380	4-1/8	4.125	3.988	4.763	4.546	1-11/16	1.744	1.632	1.151	5.750	6.000	1.250	0.072
3	3.0000	3.0000	2.9880	4-1/2	4.500	4.350	5.196	4.959	1-7/8	1.935	1.815	1.290	6.250	6.500	1.250	0.079

Source: Reprinted courtesy of The American Society of Mechanical Engineers.

Appendix B

Appendix B **207**

ANSI Hatch Patterns as Drawn with AutoCAD

Tab indicates AutoCAD hatch name. All patterns shown at 80% scale unless noted.

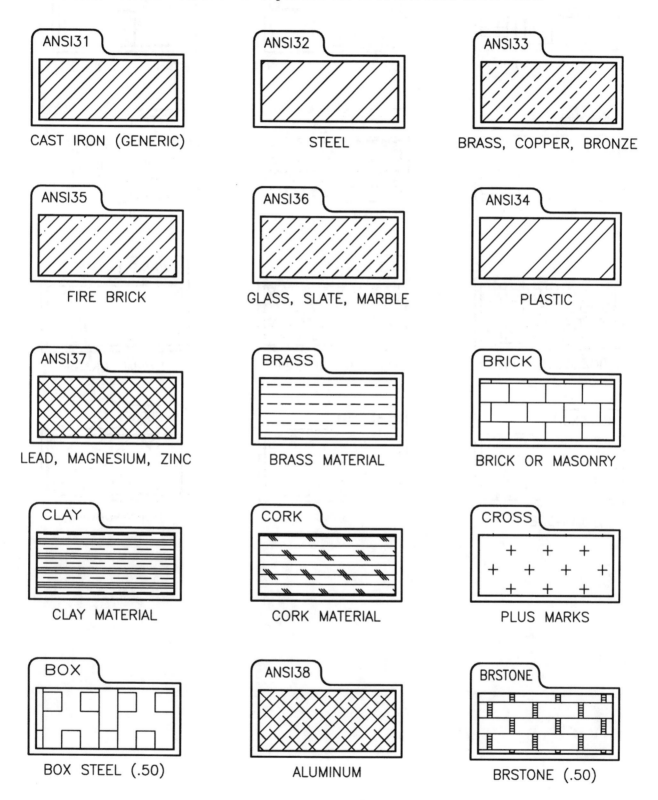

Continued on next page

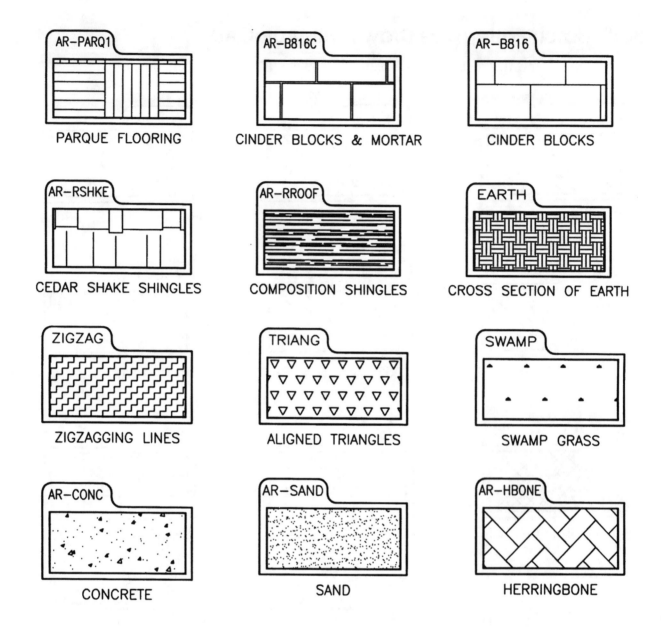

Appendix B

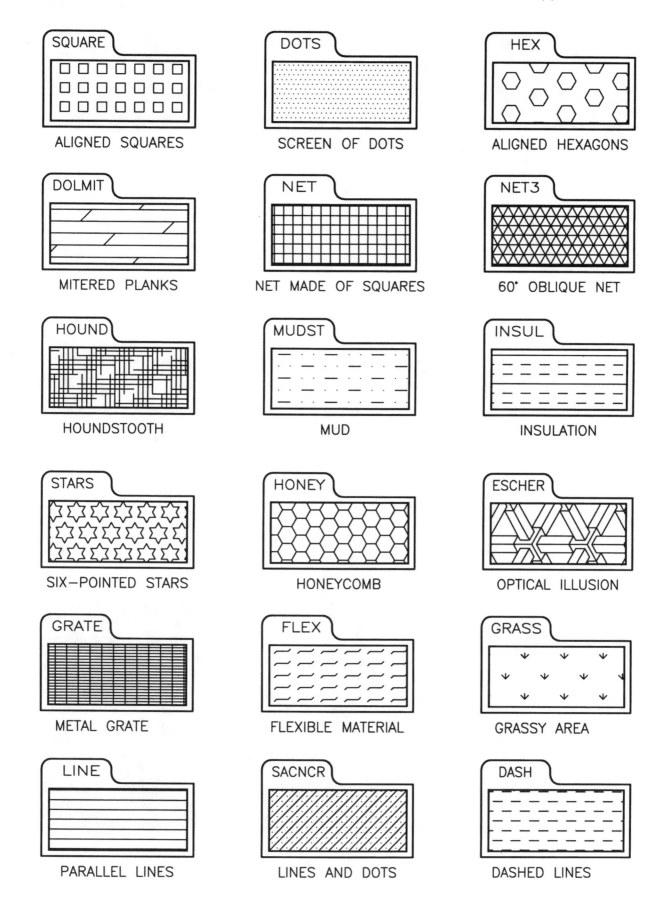

Index

A

acme threads, 96, 98f
actual size, 124
aesthetic design, 159, 160f
aligned dimensions, 74, 74f
allowance
    mating parts and, 126, 127f
alphabet
    Anglo-Saxons and, 3
    forms of, 3-4, 3f
    Gothic, 4
    Greek, 3
    Phoenician, 3
    Roman, 3
    styles of, 3-4, 4f
American National Standards Institute (ANSI, Ann-see), 4
    sheet layouts for, 146, 149f
angles
    isometric pictorial sketches of, 35, 35f
Anglo-Saxons
    alphabet and, 3
archeological digs
    graphic communications in, 3
arcs
    dimensioning of, 67, 67f
arrowheads
    on sketches, 64f, 65f
assembly drawing, 145
assembly views, 151-154, 154f-156f
    fully exploded, 156f
    partially exploded, 155f
auxiliary views, 103f, 116-117, 116f-117f
    of orthographic objects, 59f
axonometric pictorial sketch(es), 29-35, 29f
    characteristics of, 30-31
    dimetric, 30f, 32f
    isometric, 30f, 31f, 31-35
    trimetric, 30f, 32f

B

baseline (datum) dimensions, 70, 70f, 75f
basic size, 124
bilateral form, of tolerance expression, 124, 124f
bill of material (BOM), 152f
blind holes, 76f, 77f, 80f
bolt heads
    types of, 96, 99f
bolts
    screws and, 91-92
BOM (bill of material), 152f
boxes, 22f, 23
broken out sectioned views, 104, 105f

C

cabinet oblique pictorial sketches, 36, 37f
CAD. *See* computer aided drafting (CAD)
"callout box." *See* control frame, feature
carriage bolt head, 96, 99f
cavalier oblique pictorial sketches, 36, 37f
caves, prehistoric
    graphic communications in, 3
center, location of
    in perspective pictorial sketches, 40, 42f
centerlines, 48-49, 50f, 51f
chain dimensions, 70, 70f
chisel lead point, 11, 11f
circle sketching, techniques for, 25-27, 26f, 27f
circles, 21-23, 22f
    perspective pictorial sketches of, 40, 42f
clearance
    mating parts and, 126, 127f
clearance fit, 127, 128, 128f
clearance location fit, 128, 128f, 129f
close running fits, 131
close sliding fits, 131
computer aided design (CAD) software
    isometric viewpoints in, 31
computer aided drafting (CAD), 181-184
    advantages of, 182
    computer aided manufacturing, 182-183
    historical concepts in, 181
    machines in, 183
    parametric solid modeling, 183
    pencil and paper *vs.*, 182
    rapid prototypes in, 182, 183-184
    sheet layouts for, 146, 149f
    sketch finalization with, 13
    software concepts in, 181-183
    solid modeling in, 183
computer aided manufacturing (CAM), 182-183
computer numerical controlled (CNC), 183
computer-integrated manufacturing (CIM), 183
conceptual sketches, 17, 17f
cones, 23
conical lead point, 11, 11f
continuous line, 10, 12, 12f
control frame, feature
    in geometric tolerance, 135f
computer-aided engineering (CAE), 183
counterbore holes, 76f, 77f, 78f
counterdrill holes, 76f, 77f, 79f
countersink holes, 76f, 77f, 79f
crosshatch lines, 12, 13
cutting plane lines, 13
"CYA" (cover yourself always), 19
cylinders, 22f, 23
    dimensioning of, 68, 68f

D

da Vinci, Leonardo, 21, 91
dashed line, 24, 25f
datum (baseline) dimensions, 70, 70f, 75f
depth (D)
    of detailed objects, 49, 49f, 50f
design, engineering, 159–175
    aesthetic, 159, 160f
    combination in, 162
    design verification in, 162
    functional, 159, 160f
    improvement in, 162
    patent resources in, 163–164
    possibilities, exploration of, 162
    problem, identification of, 162
    process of, 159–160
    refining in, 162
    sample report of, 167–175
    selection of top three in, 162
    steps of, 160–161
    summary writing in, 162
    teams, 161–162
    testing in, 162
    to solution, 162–163
    verification of, 162
design, tolerance in, 123–139
    expression of, 124–125
    for mating parts, 126–127. *See also* mating parts, design for
    part sizes in, 124
design/manufacturing process
    sketches in, 19–20, 20f
detailed threads, 95f, 97f
dimension lines, 12
    on sketches, 63f
dimension(ing)
    "SI" symbol specifications and, 71, 71f, 85, 85f
    aligned, 71, 74f
    arcs, 67, 67f
    at "T" joints, 66, 66f
    baseline (datum), 70, 70f, 75f
    basic sizes in, 72, 72f
    between views, 68, 68f
    chain, 70, 70f
    circular view of, 67, 67f
    crossing of, 69
    cylinders, 68, 68f
    duplication of, 69, 69f
    from finished edge, 70, 70f
    in very small places, 83, 83f
    leaders in, 72, 72f
    lettering of, 5
    of large radii, 84, 84f
    of slots, 83, 83f
    of standard parts, 76–85, 76f–85f
    on inside of view, 65, 65f
    on most descriptive view, 66, 66f
    ordinate, 70, 70f
    placement of, 65, 65f, 74–75
    reference, 72, 72f
    repetitive feature, 84, 84f
    rounded corners, 67, 67f
    rules for, 65–73
    staggering of, 73, 73f
    symbols in, 71, 71f
    table driven, 75
    to hidden features, 66, 66f
    undirectional, 74, 74f
dimetric pictorial sketch, 30f, 32f
drafting paper
    freehand lettering and, 12
drafting, computer aided. *See* computer aided drafting
drawings, working, 145–156
    analysis of, 145
    assembly views and, 145, 151–154, 154f–156f
    details of, 145, 146
    layout of, 145
    parts list in, 151, 152f, 154f, 155f
    revision history in, 151
    sheet layouts in, 146–150, 147f, 148f
    specifications of, 145, 146f
    title blocks in, 150, 150f, 151
    vendor items in, 153f

E

easy rule
    for isometric pictorial sketches, 33, 33f
Edison, Thomas, 160
ellipse, diameter size of
    in isometric pictorial sketches, 33, 34f
engineers, sketching for. *See* sketch(es), 17–43
extension lines, 12
    on sketches, 63f

F

fillister bolt head, 96, 99f
finite-element analysis (FEA), 183
finite-element meshing and modeling (FEM), 183
fits. *See specific fit*
flat bolt head, 96, 99f
Ford, Henry, 160
form, in geometric tolerance, 133, 134f, 135f
fractions, height of, 5, 8f
Franklin, Benjamin, 160
free running fits, 131
freehand lettering. *See* lettering, freehand
full sectioned views, 104, 105f, 107, 107f, 108f
functional design, 159, 160f

G

general oblique pictorial sketches, 36–39, 37f
Geometric Dimensions and Tolerancing (GDT), 133
geometric tolerance, 133–139
    control frame in, feature, 135f
    form, 133, 134f, 135f
    location, 133, 134f, 135f
    orientation, 133, 134f, 135f
    profile, 133, 134f, 135f
    runout, 133, 134f, 135f
    symbol size in, 136f
geometry concepts, descriptive, 118–119
    line, point of view of, 119f
    line, true length of, 118f
    true size surfaces, 120f
"Go-Nogo" gauges, 125, 126f
Goodyear, Charles, 160
Gothic alphabet
    C. W. Reinhardt and, 4
Greek alphabet, 3
Guttenberg, Johann
    printing press and, 3–4, 4f

H

half sectioned views, 109, 1098f, 110f
hatch pattern
    in sectioned lines, 113, 113f
height (H)
    of detailed objects, 48, 49f, 50f
hex nut bolt head, 96, 99f
hexagon bolt head, 96, 99f
hidden lines, 13, 47, 47f, 48f
hieroglyphics, 3, 3f
hole types, 76f
    blind, 76f, 77f, 80f
    counterbore, 76f, 77f, 78f
    counterdrill, 76f, 77f, 79f
    countersink, 76f, 77f, 79f
    spotface, 76f, 77f, 78f
    symbols of, 77f
    tapped, 76f, 80f

I

interference fit, 127, 128f, 129
isometric pictorial sketch(es), 30f, 31f, 31–36
    disadvantages of, 31
    easy rule for, 33, 33f
    ellipse diameter size in, 33, 34f
    in computer aided design software, 31
    in parametric solid modeling packages, 31
    oblique and, 36f
    of angles, 35, 35f
    positions of, 35, 35f
    thickness in, 33, 34f

J

just in time (JIT), 183

L

laminate object machine (LOM), 184
lead, 92
    hardness of, 10–11, 11f
    points, 9–10, 10f
leaders
    on sketches, 63f
Least Material Condition (LMC), 137
lettering, freehand, 4–9
    body position in, 9
    drafting paper and, 12
    linetypes in, 9, 12–13, 12f
    of numerals, 5, 7f
    on dimensions, 5
    on notes, 4–5
    pencil type for, 9
    room humidity and, 12
    single stroke gothic, 4, 4f
    spacing rules in, 7, 7f, 8f
    tips for, 9
    tools for, 10–13
limit form, of tolerance expression, 124, 125, 125f
Lincoln, Abe, 160
line fits, 130, 130f
line sketching, techniques for, 23–24
linear spacing, 7, 8f
linetypes
    center, 52–53, 54f, 55f
    continuous, 10, 12, 12f
    hidden, 51, 51f, 52f
    noncontinuous, 10, 12–13, 12f
    using different, 51–54
    visible, 51–52, 52f, 53f
LMC (Least Material Condition), 137
location in geometric tolerance, 133, 134f, 135f
loose running fits, 131

M

mating parts, design for, 126–127
    allowance and, 126, 127f
    clearance and, 126, 127f
    fit, types of, 127–130. *See also specific fit*
Maximal Material Condition (MMC), 136
McCoy, Elijah, 160
mechanical pencil, 10
medium running fits, 131
metric thread notation, 94, 95f
milling machine, 126
MMC (Maximal Material Condition), 136
modifiers, 136, 137f

N

nominal size, 124
noncontinuous line, 10, 12–13, 12f
notes
    lettering of, 4–5
numerals
    single stroke gothic, 5, 7f

O

object lines. *See* visible lines
oblique pictorial sketches, 29f, 36–39, 37f
    cabinet, 36, 37f
    cavalier, 36, 37f
    general, 36–37, 37f
    isometric and, 36f
    orientations of, 38f
offset sectioned views, 112, 112f
Old English
    single stroke gothic and, 4, 5f
one-point perspective pictorial sketches, 39
ordinate dimensions, 75, 75f
orientation, in geometric tolerance, 133, 134f, 135f
orthographic drawing
    principal planes of, 48, 56
    principal views of, 47, 56
orthographic objects
    auxiliary views of, 59f
    interior details of, 59f
    proper alignment of, 54–55, 58f
    sketching of, 55, 56f
orthographic projection, 48f
oval bolt head, 96, 99f

P

pan bolt head, 96, 99f
parametric solid modeling (PSM), 182
parametric solid modeling (PSM) packages
    isometric viewpoints in, 31
parts list, 147, 149, 150f
patent resources
    in design, 163–164
pencil, type of
    in freehand lettering, 9
perspective pictorial sketches, 29f, 39–40
    center location of, 40, 42f
    of circles, 40, 42f
    one-point, 39
    plotting of, 40
    three-point, 39, 39f
    two-point, 39, 40f
    vanishing point in, 40, 40f, 41f
phantom lines, 13, 13f
Phoenicians alphabet, 3
pictorial sketches, 18, 19f, 23, 29–40
    axonometric, 29–35, 29f. *See also* axonometric pictorial sketches
    oblique, 29f, 36–39, 37f
    perspective, 29f, 39–40
pitch, 92, 93, 94
possibilities
    exploration of, 162
    in engineering design, 161
precision running fits, 131
printing press
    Johann Guttenberg and, 3–4, 4f
problem
    identification of, 162
    in engineering design, 161
product
    in engineering design, 161
profile, in geometric tolerance, 133, 134f, 135f

R

radii, large
    dimensioning of, 84, 84f
rapid prototypes, in computer aided drafting, 183–184
rapid prototyping (RP), 181–182
Regardless of Feature Size (RFS), 136
Reinhardt, C.W.
    Gothic alphabet and, 4
removed sectioned views, 111, 111f
revolved sectioned views, 110, 110f
RFS (Regardless of Feature Size), 136
ribs, 115, 115f
robotically guided extrusion (RGE), 184
Romans
    alphabet and, 3
round bolt head, 96, 99f
running fits
    common sizes of, 132f
runout, in geometric tolerance, 133, 134f, 135f

S

schematic threads, 95f, 97f
screw(s)
    bolts and, 91–92
    definition of, 92
    history of, 91, 91f
    parts of, 92f
sectioned lines
    aligned parts in, 114, 114f
    hatch pattern in, 113, 113f
    rules for, 113–115
sectioned views, 104–112, 105f
    broken out, 104, 105f
    full, 104, 105f, 107, 107f, 108f
    half, 109, 109f, 110f

offset, 112, 112f
removed, 111, 111f
revolved, 110, 110f
ribs in, 115, 115f
spokes in, 115, 115f
vendor parts in, 115, 115f
webs in, 115, 115f
selective assembly
    in transition fit, 129, 129f
selective laser sintering (SLS), 184
shank, 92
sheet layouts, 146–150, 147f, 148f
    size of, 146, 149f
"SI" symbol, specifications of
    dimensioning and, 71, 71f, 85f
    in title block, 151
simplified threads, 95f, 97f
single stroke gothic, 4, 4f
    Old English and, 4, 5f
    strokes in, 4, 6f
single stroke gothic numerals, 5, 7f
sketch(es), 17–43
    "CYA" and, 19
    2-D format in, 18
    archiving of, 18
    arrowheads on, 64f, 65f
    basic elements of, 20–21
    building blocks for, 21–23, 22f, 23f
    circle, techniques for, 25–27, 26f, 27f
    conceptual, 17, 17f
    design intent in, 18
    detailed, 47–59
    dimension lines on, 63f
    dimensions of, 64
    effectiveness of, 19–20
    extension lines on, 63f
    in design/manufacturing process, 19–20, 20f
    leaders on, 63f
    line, techniques for, 23–24, 24f, 25f, 26f, 27f
    of orthographic objects, 55, 56f, 57f
    ownership of, 17
    paper position for, 21, 21f
    paper type for, 21
    pencil position for, 20, 21, 22f
    pencil type for, 20–21, 22f
    pictorial, 18, 19f, 23, 29–40. See also pictorial sketches
    poor orientation of, 51f
    proper views of, 50f
    text on, 63f
    tips for, 28
sliding fits, 131
    common sizes of, 132f
slots

dimensioning of, 83, 83f
socket bolt head, 96, 99f
solid modeling
    in computer aided drafting, 183
solution
    design to, 162–163
spacing rules
    in freehand lettering, 7, 7f, 8f
spokes, 115, 115f
spotface hole, 76f, 77f, 78f
square, 21–22, 22f
square bolt head, 96, 99f
square threads, 96, 98f
standard parts
    dimensioning of, 76–85, 76f–85f
stereolithography machine (STL), 184
strokes
    in single stroke gothic, 4, 6f

T

"T" joints
    dimensions at, 6s6, 66f
table driven dimensions, 75
tapped holes, 76f, 80f
text
    on sketches, 63f
threads. See also screws
    acme, 96, 98f
    axis, 92
    detailed, 95f, 97f
    internal, 96, 97f
    major diameter of, 93, 94
    metric notation of, 94, 95f
    nomenciature of, 93f
    note for U.S. screw, 93, 94f
    pitch of, 93, 94
    schematic, 95f, 97f
    simplified, 95f, 97f
    square, 96, 98f
    symbols, 94–96, 95f
three-point perspective pictorial sketches, 39, 39f
title blocks, 150–151, 149f, 150f
tolerance(ing)
    definition of, 123, 131
    geometric, 133–139
    summary of, 131
tools, in freehand lettering, 10–13
    lead points, 10–12, 10f
    pencil lead hardness in, 11, 11f
transition fit, 127, 129, 129f
    selective assembly in, 129, 129f
triangle, 21–23, 22f
trimetric pictorial sketch, 30f, 32f

2-D format
    in designs, 47, 47*f*
    in sketching, 18
two-point perspective pictorial sketches, 39, 40*f*

U

undirectional dimensions, 74, 74*f*
unilateral form, of tolerance expression, 124, 125, 125*f*

V

vanishing point
    in perspective pictorial sketches, 39, 40*f*
vendor items, 153*f*
vendor parts, 115, 115*f*
visible lines, 12, 51–52, 52*f*, 53*f*, 109
volume, letter spacing by, 7, 8*f*

W

webs, 115, 115*f*
wedge lead point, 10, 10*f*
wedges, 22*f*, 23
Whitney, Eli, 91, 92*f*, 160
    interchangeable parts and, 126
width (W)
    of detailed objects, 48–49, 49*f*, 50*f*
World Wide Web (WWW), 92
Wright, Orville and Wilbur, 160